INDOOR POLLUTION

A Reference Handbook

Other Titles in ABC-CLIO's
Contemporary
World Issues
Series

Biodiversity, Anne Becher
Crime in America, Jennifer L. Durham
Environmental Justice, David E. Newton
Legalized Gambling, second edition, William N. Thompson
The Ozone Dilemma, David E. Newton
Recycling in America, second edition, Debi Strong
School Violence, Deborah L. Kopka
Victims' Rights, Leigh Glenn
Welfare Reform, Mary Ellen Hombs
Women in the Third World, Karen L. Kinnear
United States Immigration, E. Willard Miller and Ruby M. Miller

Books in the Contemporary World Issues series address vital issues in today's society such as terrorism, sexual harassment, homelessness, AIDS, gambling, animal rights, and air pollution. Written by professional writers, scholars, and nonacademic experts, these books are authoritative, clearly written, up-to-date, and objective. They provide a good starting point for research by high school and college students, scholars, and general readers as well as by legislators, businesspeople, activists, and others.

Each book, carefully organized and easy to use, contains an overview of the subject, a detailed chronology, biographical sketches, facts and data and/or documents and other primary-source material, a directory of organizations and agencies, annotated lists of print and nonprint resources, a glossary, and an index.

Readers of books in the Contemporary World Issues series will find the information they need in order to better understand the social, political, environmental, and economic issues facing the world today.

INDOOR POLLUTION

A Reference Handbook

E. Willard Miller
Department of Geography

Ruby M. Miller
Pattee Library

The Pennsylvania State University

Santa Barbara, California
Denver, Colorado
Oxford, England

Library of Congress Cataloging-in-Publication Data
Miller, E. Willard (Eugene Willard), 1915–
Indoor pollution : a reference handbook / E. Willard Miller and Ruby M. Miller.
p. cm. — (Contemporary world issues)
Includes bibliographical references and index.
ISBN 0-87436-895-2 (alk. paper)
1. Indoor air pollution—Handbooks, manuals, etc. I. Miller, Ruby M. II. Title. III. Series.
TD883.M52 1998
363.739'2—dc21 98-25606
CIP

04 03 02 01 00 99 10 9 8 7 6 5 4 3 2

ABC-CLIO, Inc.
130 Cremona Drive, P.O. Box 1911
Santa Barbara, California 93116-1911

Typesetting by Letra Libre

This book is printed on acid-free paper ♾.
Manufactured in the United States of America

Contents

Preface

The presence of polluted air in the home from heating and cooking food is as old as civilization itself. In modern times the general public only became aware of the health hazards when the number of household chemicals increased dramatically in the twentieth century. Few governmental regulations have been passed to regulate the home environment. The concept that the home is a man's castle has delayed the passing of governmental regulations that control the quality of the indoor environment. Standards are, however, being established by many private organizations.

This volume begins with a perspective on indoor pollution. Initially it provides a background on the factors affecting indoor air quality, health effects, development standards, and the difficulties of developing governmental regulations. This is followed by a discussion of the sources of indoor pollution including ambient air pollutants, bioaerosols, such physical pollutants as radon, asbestos, heavy metals, and parcticulates, volatile organic compounds such as formaldehyde, polynuclear aromatic compounds, polychlorinated biphenyls, and pesticides, and concludes with a discussion on noise pollution. A major section treats sick building syndrome and building-related illnesses such as Pontiac fever, *Legionella pneumophila* fever, rhinitis, humidifier fever, and hypersensitivity pneumonitis. The initial chapter concludes with the means of controlling such indoor pollutants as particulate contaminants, gaseous combustion products, gas phase contaminants, volatile organic compounds, and radon.

Federal legislation that directly considers indoor pollution in buildings is limited to only radon, lead paint, and asbestos. A number of related laws and regulations include tobacco's effect on health, energy conservation and sick building syndrome, occupational safety and health, toxic substances, clean air, and noise.

Many organizations, although broad in scope, consider problems of indoor pollution. The principal federal government agencies are listed as well as how they treat indoor pollution problems. Many private organizations, such as the Adhesive and Sealant Council, Alliance to End Childhood Lead Poisoning, and the Cure Formaldehyde Poisoning Association, have established standards to control indoor pollution.

There is a considerable body of literature on indoor pollution. The studies are increasing particularly as to how indoor pollution can be controlled. The literature varies from scientific to socially oriented studies. The bibliography provides a list of books, journal articles, and governmental documents as well as the leading journal sources.

There is considerable audiovisual material available on indoor pollutants. These include films on ozone, radon, tobacco smoking, food poisoning, asbestos, chemical toxins, and noise. The volume concludes with such appendixes as a glossary, organizations, measurements, chemical compounds, testing information, and sources of volatile organic compounds in indoor air.

E. Willard Miller
Ruby M. Miller
Pennsylvania State University

Indoor Pollution: A Perspective

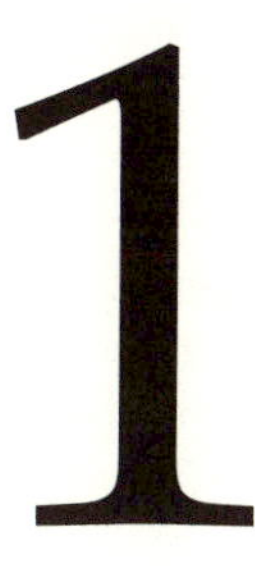

Problems of indoor air quality began when man first brought fires into a cave, but only in the past several decades has the general public become aware of the health hazards of contaminated air. In the early 1970s, the presence of friable asbestos in the atmospheres of public schools and other buildings created an awareness of atmospheric health problems. In response, the federal government passed legislation to remove asbestos from public buildings.

During the same period the energy crisis, beginning in 1973, exacerbated the indoor pollution problem. Older and newer buildings were more highly insulated in order to conserve energy. As a result, the exchange of outdoor-indoor air in buildings greatly lessened, and indoor pollution built up. Further, pollutants brought in from the outdoors added to those already present. Hundreds of outbreaks of illnesses among the occupants of new and recently remodeled offices, schools, homes, and other buildings began to be reported, creating what now is known as sick building syndrome.

There were other health warnings in the late 1970s and early 1980s due to poor air quality. There was an outbreak of Legionnaires' disease (pulmonary lung infection) in Philadelphia in 1976. Numerous complaints of odors and acute irritating symptoms were received from owners of houses insulated with urea-formaldehyde foam. It also became evident that radon contamination of homes was not associated only with uranium mill waste tailings and abandoned phosphate strip mines but was a worldwide problem affecting millions of homesites.

Although the U.S. government has set standards for outdoor particulate matter, carbon monoxide (CO), ozone (O_3), nitrogen dioxide (NO_2), sulfur dioxide (SO_2), lead, and nonmethane hydrocarbons, it has been reluctant to provide guidelines for indoor pollution. It is considered a delicate and difficult matter to establish rules that apply to the home environment. As a result, the United States currently has no legally enforceable, health-related national standards for human living space. Consequently, there has been a lack of developing governmental regulation by such agencies as the U.S. Environmental Protection Agency (EPA) for air quality in homes.

Indoor Air Quality

Indoor air quality (IAQ) is a complex issue. Studies indicate that more than 900 possible contaminants—from thousands of sources—are present in a given indoor environment. The creation of the indoor environment in any building involves the interactions of a complex set of factors that are constantly changing.

In 1987, the National Institute of Occupational Safety and Health (NIOSH) conducted a survey of 529 buildings whose occupants complained about IAQ. This study revealed that poor IAQ was due to:

Inadequate ventilation	53 percent
Outside contamination	10 percent
Microbial contamination	5 percent
Building material contamination	4 percent
Unknown	13 percent

It was evident that inadequate ventilation was a major problem in buildings. As a result, the past ventilation practices were considered totally inadequate and a major cause of poor IAQ. Continued study of ventilation has indicated that increased ventilation is a means of diluting airborne environments. However, the study also recognized that inadequate ventilation was not the only contributor to poor IAQ. The quality of indoor air is due to many factors. Further, in any specific building any one of the several factors may dominate.

Pollutant Characteristics

The characteristics of a pollutant are important in determining indoor concentrations. For example, carbon monoxide and nitrogen dioxide are both products of combustion, but their behavior in the atmosphere is quite different. In the indoor atmosphere, the reduction of the NO_2 concentration is due not only to air moving out of the building but also to the chemical reaction that is experienced by NO_2. In contrast, CO is an inert gas that is not changed by indoor chemical reactions. Consequently, indoor concentrations of NO_2 decrease more rapidly than CO concentrations.

Pollutants of lighter weights disperse faster as air movement increases inside the building.

Factors Affecting Indoor Air Quality

The quality of indoor air is affected by many factors. Most pollutants of indoor air originate inside buildings. Outdoor air may enter buildings and add to pollution. Due to the huge variety of possible pollutants, the problem of air quality can be extremely complex.

Indoor Sources

Indoor pollution comes from many sources, and amounts vary depending upon indoor conditions. For example, when temperature or humidity increases, the concentration of pollutants will normally increase. Pollutant levels may also increase or decrease due to indoor functions that change during the day (e.g., a gas range might operate only certain times throughout the day). Thus, the variations in pollutant emissions, either due to environmental changes or changes in operational practices, must be evaluated.

Outdoor Concentrations

Outdoor air will flow into buildings with changes in barometric pressure and temperature. The concentration of pollutants in the outdoor air determines their effect on the quality of the indoor air. In a dry region, the outdoor air may contain a large quantity of particles (dust). In the center of a city, the outdoor concentration of CO can vary from 1 to 9 milligrams per cubic meter (mg/m^3), primarily due to motor vehicle traffic. In addition, the indoor CO concentration peaks later in the day than does the outdoor concentration, due to a general buildup indoors. Thus, the building environment has a delaying effect on the indoor peak concentration. This effect is only temporary, for if the outdoor concentration remains high, the indoor concentration will eventually equal that of the outdoor concentration.

Volume

The indoor concentration of pollutants will depend upon the amount of pollutants present in a given area. The volume available for pollutants in a building varies from that of the entire building to a small portion of the structure. The same amount of pollutants emitted in an area of 100 m^3 compared to an area of 200 m^3 will result in different concentrations.

In most structures, not all air volume is equally accessible for pollutants to disperse. This is easily recognizable if areas are blocked by physical restraints, such as a door. However, the dispersal of pollutants is sometimes blocked by invisible barriers (e.g., if an indoor combustion source is located on an upper floor, the pollutants concentration would be higher on the upper floor and lower on lower floors). This tendency is due to thermal buoyancy, which restricts dispersal to lower floors. Similarly, if a source of pollutants is on a lower floor, the concentration elsewhere in the structure depends on air circulation among the floors.

The use of fans is an important factor in detecting the volume of a pollutant in a given place. For example, if radon enters a building through the basement, the concentration will be high on the lower floors. If, however, a circulating fan is operating, the concentration on upper floors will increase; at the same time the concentration will decrease in the basement.

Exchange Rate

Air exchange determines the rate at which indoor concentrations of pollutants will decrease due to the importation of outdoor air. The rate is expressed as the volume of air exchanged per unit of time. The volume of air is commonly expressed in terms of the total volume of the structure; one hour is normally used as the time unit. For a house with a 200 m^3 volume where 50 m^3 of outdoor air enters per hour replacing indoor air, the air exchange rate is 0.25 air changes per hour (ACH).

The rate of air exchange determines the amount of time required for the indoor pollutant concentration to decrease. For example, at 0.8 ACH, pollutant concentration is reduced to half its peak value in about one hour. In contrast, at 0.2 ACH, the same reduction takes four times longer.

The air exchange rate has three major elements: infiltration, natural ventilation, and mechanical ventilation. A pressure difference between the indoors and outdoors is responsible for infiltration of air into a building. Natural ventilation refers to air flowing into and out of a structure through open windows and doors. Mechanical ventilation refers to air exchange that is driven by a motorized system of fans. Local or spot ventilation refers to the ventilation of only a part of a building, such as a system that removes air from a bathroom.

To create an infiltration of air into a building, temperature differences between indoor and outdoor air are of major importance. In the winter, warm air rises and escapes through openings at the higher levels, whereas cold air enters through openings in the lower parts of the building.

Air exchange rates vary considerably over time. They can change hour to hour on any given day and can vary substantially season to season. These variations are due to changes in many factors.

Health Effects of Air Pollutants

The major reason to measure and control indoor air pollutants is to determine their effect on human health. To acquire this information, it is necessary to measure not only the ambient air concentration of pollutants but also how long the individual has been breathing the pollutant. In addition, how an air pollutant interacts with the human body provides the most useful information about the dose to a target organ or bodily system. The human body has a tremendous capacity to absorb all types of chemicals and either utilize them to support some bodily function or eliminate them, but when the pollutant achieves a level in the body beyond the capacity to remove, then it becomes toxic. This is called beyond total body burden.

The concept of total body burden refers to the amount of toxic material that collects in a human system. Pollutants, such as lead, formaldehyde, and others, are accumulated initially in the blood, urine, soft tissue, hair, teeth, and bone. They are then removed by normal body processes. The time it takes the normal body processes to remove trace materials from the human system may vary from a few days to several months or sometimes even years.

Within the general population, certain groups sensitive to specific air pollutants have been identified. They include very young children, whose respiratory and circulatory systems are still developing, the elderly, whose respiratory systems function poorly, and persons who have certain diseases such as asthma, emphysema, and heart disease.

Air pollution principally affects the respiratory, circulatory, and olfactory systems. The respiratory system is the principal route of entry of pollutants into the body, some of which may alter the function of the lungs. For example, carbon monoxide enters the body by inhalation and is absorbed directly into the bloodstream. The total body burden is thus located in the circulatory system, of which the lungs dominate.

Standards for Indoor Air Quality

Although there has been minimal federal legislation controlling indoor air quality, a number of engineering societies have established guidelines. Because of the recognized importance of ventilation, the American Society of Heating, Refrigerating, and Air-Conditioning Engineers (ASHRAE) has provided a set of indoor ventilation standards (ASHRAE 62-1989). These are now widely accepted. Most U.S. heating, ventilation, and air-conditioning design firms now use these standards for minimum air design requirements for commercial buildings. Many building codes, such as those designed by municipalities, have officially incorporated ASHRAE 62-1989 into their design standards, thereby making these ventilation standards an enforceable minimum requirement. The ASHRAE design engineers recognize the importance of ventilation control for two reasons: Ventilating air is a means of controlling indoor air contamination through the amount of admission of fresh outdoor air, and control of ventilation has an impact on the energy consumption in the building (see Table 1).

TABLE 1
Federal Air Quality Standards (μg/m³)

Time	SO_2	NO_2	CO	Particulate Matter
Primary Standards				
24-hr. maximum	365	—	—	260
8-hr. maximum	—	—	10,000	—
3-hr. maximum	—	—	—	—
1-hr. maximum	—	—	40,000	—
Secondary Standards (when different from primary standards)				
24-hr. maximum	260	—	—	130
3-hr. maximum	1,300	—	—	—

Source: Clean Air Amendments of 1970, Public Law 91-604, 84 Statute, December 31, 1970. Air Standards Authorization implemented by U.S. Environmental Protection Agency, Washington, DC, April 30, 1971.

ASHRAE 62-1989 has established two procedures for determining an acceptable ventilation rate. The Ventilation Rate Procedure specifies a minimum ventilation rate based upon the space functions within given building types. The Ventilation Rate Procedure tables are derived from respiration rates, which are based on the occupants' activities.

The second procedure based on ASHRAE 62-1989 is called the Indoor Air Quality Procedure. This procedure requires monitoring certain indoor air contaminants below specified values. The contaminants and their levels are based upon concentrations determined by the EPA and the American Conference of Governmental Industrial Hygienists (ACGIH). The maintenance of the contaminant level below specified values may be achieved through source control, the use of air cleaning, or local exhaust ventilation. Occasionally, the minimum ventilation rate of 74 cubic feet per minute (cfm) per person required in the Ventilation Rate Procedure may be reduced by utilizing the Indoor Air Quality Procedure. In order for this procedure to be effective the amount of contaminants in the outdoor air must be determined. The EPA has set acceptable levels of certain contaminants for outdoor air. Other contaminants may be available at the building site, where, by increasing the contaminants in the outdoor air for ventilation, the heating, ventilation, and air-condition (HVAC) system becomes a pathway through which those contaminants can travel into the indoor environment.

Civil Litigation

Increased awareness of the health risks of air pollution and the fact there is little governmental regulation combine to make indoor air pollution a fertile field for civil litigation. Individuals who claim that their health has been harmed by air pollution have launched lawsuits against employers, manufacturers of chemicals and building products, contractors, architects, and real estate professionals. Many of these cases turn on interpretations of common law principles such as negligence, caveat emptor, and others. Principles of tort law, which can apply to private industry in any number of cases, are important. Many of the allegations of injury can also spring from contractual relationships, and so contract law can be a basis for legal action.

In recent years indoor pollution has become an important topic for the media to report to the general public. Print and broadcast media often report the problems of air pollution in graphic detail. So-called sick building episodes, in which individuals experience such problems as coughing, headaches, dizziness, nausea, and respiratory irritation, are often reported. When these conditions are reported to building owners and operators, they may respond in many ways. If, however, the individuals in a building do not receive what they consider to be a satisfactory response, they are quick to begin legal action.

In considering the recovery of damages, two objectives must be recognized. First, compensation is made to those innocently injured by the unlawful behavior of others. The term *unlawful* does not mean *illegal* but simply involves a lack of care for the welfare of others due to breach of terms in a contract. Our justice system is based on the concept that the innocent person who has been injured has some recourse for the improper actions of others.

The second objective is deterrence. Common law requires that those who have unlawfully caused injury must be punished, which in turn provides incentive to avoid the same situation in the future. The objective in this context is to control, by relying on financial incentive rather than the law of regulations. Thus, common law provides a means of controlling and preventing new indoor air pollution.

The legal cases brought under common law alleging that health has been harmed by air pollution have been extremely complex. A litany of factors, such as improper ventilation, bacteria, or other microorganisms in ventilating systems, ventilation of automotive pollutants from the outdoor air into the building, chemicals used in the building, and materials used in the building, make it very difficult for fact finders to determine who is responsible for an indoor air quality problem. Plaintiffs' lawyers have had to be creative in their approach. If uncertain whom to sue, they sometimes file suit against many defendants and construct novel chains of causation.

For plaintiffs, these lawsuits present formidable barriers. To begin with, the suit must be filed within the applicable statute of limitations. They must prove that the defendant(s) caused the injury and were negligent in so doing. Procedural requirements, as well as the rules of evidence, will also complicate the plaintiff's task. They must also find means of financing the litigation, which is likely to be very costly, and face lengthy delays on appeal before receiving any compensation. Economic damage, in the form of reduced property value or cleanup costs, may be recovered. Plaintiffs can also recover damages for health harm, although causation is more difficult to prove, and the characteristics of some diseases such as cancer make it extremely difficult to recover expenses for the most severe health risks of indoor air contamination.

Defendants also face many problems. Costs run high, because evidence is likely to be complex and require technical expertise. In addition, because of the cost of defense, nuisance suits can coerce a settlement. In the event of an adverse defense judgment, the losses can be staggering, sometimes millions of dollars.

The number of civil cases has been relatively small but is growing rapidly. The most common and successful lawsuits have involved relatively minor asbestos conditions. In contrast, lead and radon contaminations have produced little litigation. A major problem is to provide proof of a problem within the statute of limitations. For example, it is easier to recover damages for minor irritant injuries than for diseases that require decades to develop, such as cancer. The litigation system is likely to produce inconsistent results and offer no permanent solution for indoor air pollution problems. In the face of such uncertainty, defendants can protect themselves by buying insurance against contamination problems creating health hazards, thus minimizing risk of a costly lawsuit.

Sources of Indoor Pollutants

Ambient Air Pollutants

Ambient (outside) air pollutants include carbon dioxide (CO_2), carbon monoxide, nitrogen dioxide, sulfur dioxide, ozone, and passive (environmental) tobacco smoke. Federal air quality standards have been established for air pollutants (see Table 2).

Passive Tobacco Smoke

It is now recognized that passive (environmental) tobacco smoke found indoors can produce measurable physiological changes of human tissue, resulting in a diseased condition. Ambient tobacco smoke is a recognized carcinogen associated with an increased risk of lung cancer.

Environmental tobacco smoke consists of a mixture of solids, liquids, and dissolved gases. It is the most important source of indoor air particulates other than house dust. The composition of tobacco smoke is extremely complex. Approximately 4,700 different types of chemicals have been identified.

During the smoking process, contaminants include sulfur dioxide, ammonia, nitrogen oxides, vinyl chloride, hydrogen

TABLE 2
Hazardous Air Pollution Levels

Air Pollutants	Concentration	Times Duration (average)
Sulfur dioxide	2,620 μg/m³	24 hrs.
Carbon monoxide	57.5 mg/m³	8 hrs.
	86.3 mg/m³	2 hrs.
	1,400 mg/m³	1 hr.
Nitrogen dioxide	3,750 μg/m³	1 hr.
	938 μg/m³	2 hrs.
Particulate matter	1,000 μg/m³	24 hrs.
Ozone (photochemical oxidants)	800 μg/m³	4 hrs.
	1,400 μg/m³	1 hr.

Source: Clean Air Amendments of 1970, Public Law 91-604, 84 Statute, December 31, 1970. Air Standards Authorization implemented by U.S. Environmental Protection Agency, Washington, DC, April 30, 1971.

cyanide, formaldehyde, radio nuclides, benzene, arsenic, and many others. Tobacco smoke is the most important source of benzene, a known carcinogen, which produces tumors, in the home. Benzene levels are 30–50 percent higher in homes with smokers than in nonsmoking households. Of the chemical products from smoking, carbon monoxide, nicotine, and tar particulates are known to have a detrimental impact on health. There are more than 40 carcinogenic compounds in tobacco smoke. Further, tobacco smoke is a source of mutagenic substances, that is, compounds that cause permanent change in the genetic structure of cells.

Until the 1950s, the public perceived smoking as a pleasant experience, conducive to a social environment. Tobacco advertisements gave smoking an aura of well-being. The first evidence that smoking could be injurious to health came in 1950 in a report published in the *Journal of the American Medical Association* (Rosen and Levy, 1950). The study described a case where a baby developed severe asthma. When the mother stopped smoking, the baby's symptoms disappeared. Eighteen months later the baby's symptoms returned when the mother resumed smoking. The next research study on the effects of tobacco smoke did not appear until the 1970s. Since then, hundreds of studies have been undertaken on the effects of smoking on individuals and the effects of breathing passive smoke.

It is now recognized that cigarette, cigar, and pipe smoke are significant sources of air pollution that can harm nonsmokers. This passive smoke released into the ambient environment (so-called secondhand smoke) contains the same gaseous and particulate pollutants as those received by the smoker. The passive smoke released into the indoor air frequently has higher concentrations of pollutants than that received by the smoker. This results when a smoker does not inhale.

A nonsmoker's actual exposure to tobacco smoke depends on many variables. These include where the nonsmoker is located with respect to a smoker, the extent and form of ventilation, and the number of cigarettes smoked within a set period of time. There are now attempts to control indoor smoke pollution by prohibiting the use of tobacco in certain places such as restaurants. The effects of tobacco smoke in indoor space are directly proportional to the smoke density and inversely proportional to the effective ventilation rate.

Since the early 1980s, studies have related involuntary smoking to increased risk of developing cancer (see Table 3).

TABLE 3
Annual U.S. Deaths and Cases from Secondhand Tobacco Smoke

Disease	Number of Deaths or Cases
Lung cancer	3,000 deaths
Heart disease	35,000–62,000 deaths
Sudden infant death syndrome	1,900–2,700 deaths
Low–birth weight babies	9,700–18,600 deaths
Asthma in children	8,000–26,000 new cases
	400,000–1 million cases of existing disease, getting worse
Bronchitis in children	150,000–300,000 cases
	7,500–15,000 hospitalizations
	136–212 deaths

Source: California Environmental Protection Agency.

One of the earliest studies took place in Japan when 91,504 nonsmoking Japanese women were surveyed for a period of 14 years (1966–1979) to reveal the effects of involuntary smoking on the lungs. All of the husbands smoked indoors. Of the study group, 174 women died of lung cancer. The study revealed that the nonsmoking wives of men who smoked more than 20 cigarettes a day had 2.4 times the risk of developing lung cancer compared to wives of nonsmoking husbands. It was also found that there was an even higher risk of secondhand smoke exposure among wives of smoking agricultural workers (about 4.6 times the risk of women married to nonsmokers). There were similar, smaller increases in risk for women married to light smokers. The risks of passive smoke in an urban environment appear to be somewhat lower, but this may be due to the existence of other carcinogens, which make it more difficult to detect the effects of tobacco smoke (Hirayama, 1983).

There have been a number of studies on the effects of smoke on general health. It is now accepted that passive smoking substantially increases respiratory illness in children. Children with smoking parents have increased incidences of respiratory tract infection, bronchitis, asthma, and pneumonia. Additional effects on children include increases in coughing, wheezing, and sputum production, slower lung function growth, and more colds. These health problems increase with the number of smokers in the house (Cameron et al., 1969; Colley et al., 1974; Tager et al., 1979).

There is strong evidence that environmental tobacco smoke can aggravate the health problems of people with heart and lung diseases and allergies. A number of studies have shown that passive smoking increases the heart rate, blood pressure, and carboxyhemoglobin (COHb) and reduces the time a person can exercise without strain and chest pain (Aronow, 1977).

As a response to the growing evidence that passive tobacco smoke is harmful to health, legal actions are now beginning to appear. In 1997, an airline flight attendant diagnosed with cancer claimed that the disease was acquired due to many years of working in airplane cabins where smoking was permitted. A class action suit has been filed in Florida against the airline for damage of health.

The control of tobacco smoke in the home is difficult. Although normal ventilation will reduce the concentration of some smoke constituents, their effective removal requires unrealistic ventilation rates. This is particularly the case in cold regimes, where heat is also removed with the contaminants. The best way to control involuntary smoke is to provide separate areas for smokers and nonsmokers. Many organizations now prohibit smoking throughout the entire building.

Carbon Monoxide

Carbon monoxide is a colorless, odorless, and tasteless gas. It occurs when there is incomplete combustion of carbon-containing material. Primary sources include cooking and heating appliances, motor vehicles running in closed garages, tobacco smoke, and wood-burning fireplaces or stoves (see Table 4).

If a house is highly insulated to reduce the escape of heat, there is a danger that CO will build up to unsuspected and unhealthy—even deadly—levels.

Carbon monoxide in the home's atmosphere can be detected by an alarm, somewhat similar to a smoke detector, that meets the Underwriters Laboratories standard (UL2034, October 1, 1995). CO detectors are produced in electric- and battery-powered models. The U.S. Common Safety Commission recommends installing at least one CO detector near sleeping areas, perhaps next to the smoke detector.

A number of health effects have been associated with exposure to carbon monoxide. The severity of the health problem depends upon the increase of CO in the bloodstream. Carbon monoxide reacts with hemoglobin in blood to form carboxyhemoglobin. As a result of this interaction, the blood's ability to carry oxygen (O_2) to the body tissues is reduced.

TABLE 4
Environmental Indoor Concentration of Carbon Monoxide

Location	Mean CO Concentration (ppm)
Public garage	13.46
Service station	9.17
Other repair shop	5.64
Shopping mall	4.90
Residential garage	4.35
Restaurant	3.71
Office	3.59
Auditorium, sports arena, concert hall, etc.	3.37
Store	3.23
Health care facility	2.22
Other public buildings	2.15
Manufacturing facility	2.04
Residences	2.04
School	1.64
Church	1.56

Source: Akland, G. G., T. D. Hartwell, T. R. Johnson, and B. W. Whitmore. "Measuring Human Exposure to Carbon Monoxide in Washington, DC, and Denver, CO, during the Winter of 1982–1983." *Environmental Science & Technology* 19 (1985): 911–918. Copyright © 1985 American Chemical Company. Reprinted with permission from *Envir. Sci. Technol.*

Normal metabolic processes in the body will result in a COHb level among nonsmokers of 1.2–1.5 percent. This is a level of 10–15 parts per million (ppm) of CO. For smokers, it rises to about 3–4 percent.

At levels below 2 percent of COHb, no health problems have been found. When it rises above 2 percent COHb, the early evidence of CO poisoning includes headache, fatigue, shortness of breath, dizziness, and nausea. These symptoms can frequently be misdiagnosed as "a touch of the flu." With levels of 5 percent COHb, which is about 30 ppm CO in the atmosphere, symptoms worsen with vomiting, confusion, and heart palpitations. When the COHb level reaches 10 percent or greater there are cardiovascular and neurobehavioral problems. The final phase of poisoning results in coma, with death resulting from asphyxiation. Many people are asleep when overcome by CO gas or fall asleep as a result of the poisoning and never wake up. Every year nearly 600 Americans die from accidental carbon monoxide poisoning at home (EPA and Consumer Product Safety Commission, 1993).

Ozone

Ozone in the atmosphere continues to be a major environmental problem. Unlike most air pollutants, ozone is considered a secondary air pollutant produced in the atmosphere by photochemical processes. Photochemical production of ozone in the surface atmosphere depends upon the presence of primary pollutants such as NO_2, aldehydes, carbon dioxide, carbon monoxide, and other radicals, along with sunlight and suitable temperatures. Through the photolytic cycle, nitric oxide (NO), emitted by stationary and mobile sources, is oxidized and converted to nitrogen dioxide. The photolysis of NO_2 by sunlight thus converts it back to NO, and the free oxygen atom combines with the oxygen atom to form ozone. At the same time, the NO formed through the photolysis of NO_2 is also an O_3 scavenger. Thus, nitric oxide contributes to production and destruction of ozone.

In addition to the photochemical production of O_3 in the lower atmosphere, the level of O_3 concentration at the earth's surface is affected by a movement downward from the stratosphere to the troposphere. Ozone is created in the stratosphere by the photolysis of molecular oxygen. The downward diffusion of ozone may occur through convection, such as along low pressure troughs, during episodes of frontal passage or through jet stream interactions.

Ozone reaches its maximum concentration in the atmosphere early in the afternoon, when the peak concentration is about 350 $\mu g/m^3$. Mean hourly ozone concentration may exceed values of 240 $\mu g/m^3$ for ten hours or more. Since the early 1920s, studies have revealed that outdoor ozone is the major source for that found inside buildings. The concentration indoors is thus directly related to the outside concentration.

A number of studies have documented that activated charcoal filters can significantly reduce indoor ozone concentrations in buildings with mechanical ventilation systems. In buildings with natural ventilation, occupants can be encouraged to open windows during the cooler part of the day, normally nights and early morning, and to close windows during the warmest part of the day. This reduces the ventilation rates during the period when ozone concentrations reach the highest levels outdoors.

Although most ozone is generated outdoors, there are also indoor sources. Ozone is produced indoors from copy machines, laser printers, ultraviolet lights, and electronic devices such as air cleaners. In electronic gadgets, for example, the electrons present in the air become accelerated and positively charged ions (atoms). These liberated ions strike O_2 molecules, producing more positive ions, which become vast in numbers. Because of the large amount of energy created, the atoms combine with O_2, producing ozone.

Ozone was long ignored as a hazardous indoor air pollutant. It is now recognized that low concentrations of O_3 are highly toxic and affect health. Ozone is a powerful oxidant and as such reacts virtually with every class of biological substance. As a consequence, all indoor devices that produce ozone are regulated for O_3 output by the Food and Drug Administration (FDA). The FDA ozone limit for electronic air cleaners and other electronic devices is 0.5 ppm. Any device that exceeds this output of ozone cannot be used indoors.

At low concentrations ozone irritates the lung tissues. As a result, there can be a significant impairment of pulmonary functions, revealed by respiratory symptoms. When hourly average concentration levels exceed 200 $\mu g/m^3$, studies report eye, nose, and throat irritation, chest discomfort, coughing, and headaches (EPA, 1986). A decrease in pulmonary function in children has been reported when the hourly average of O_3 is in the range of 160–300 $\mu g/m^3$ (Lioy et al., 1985). Athletic performances are decreased when O_3 levels are in the range of 240–740 $\mu g/m^3$ (Wayne et al., 1967). There is also evidence that the incidence of asthmatic attacks are increased at similar levels of ozone. One study at New York University suggests that cumulative exposure to ozone causes loss of lung function. A study by Lippmann (1989) indicates that chronic exposure to ozone may result in accelerated aging of humans.

Carbon Dioxide

Carbon dioxide is a colorless, odorless gas. It exists in minute quantities in normal air. Indoors it accumulates as humans continuously emit CO_2 while breathing. Carbon dioxide is the principal combustion product when burning fuels. Gas, kerosene, and wood-fueled appliances are major sources of CO_2.

The amount of CO_2 accumulating indoors varies considerably. The amount of CO_2 normally exhaled by an adult in a room is about 200 milliliters per minute (ml/mn). In homes with no fuel emissions of CO_2, concentrations normally range from 0.07 to 0.24 percent of the atmosphere (National Research Council,

1981). Concentrations when fuels are burned can range from 1 to 9 percent.

Studies by the U.S. Consumer Safety Products Commission indicate that when the concentration of CO_2 in the atmosphere is below 1 percent there is no health effect. Prolonged exposure of healthy individuals to 1.5 percent CO_2 apparently causes mild metabolic stress. Respiration is affected, and breathing becomes faster and more difficult. When the CO_2 level reaches 7–10 percent in the atmosphere, human exposure produces unconsciousness within minutes, resulting in death.

There are only a few studies revealing the effect of CO_2 on human health. One major study considers the effects of CO_2 exposure on a crew on a nuclear submarine to concentrations of CO_2 of 0.7–1.0 percent. The study revealed a consistent increase in respiratory volume and cyclic changes in the acid-base balance in blood. Further study may reveal that these changes may cause reduction in bone density due to release in calcium (Schaefer, 1979).

In order to reduce CO_2 content of indoor air, ventilation guidelines have been established by ASHRAE. According to its guidelines (ASHRAE Standard 62-1981), sufficient outdoor air must be provided per person to ensure that metabolically augmented CO_2 levels do not exceed 2,500 ppm.

These guidelines do not carry the force of law unless included in state or local building codes. They are, however, widely accepted by architects in order to provide occupants of buildings a reasonably comfortable environment. These guidelines have been frequently compromised by the need to conserve energy. The ASHRAE provisions for ventilation from the outside environment impose significant heating and cooling costs.

Nitrogen Dioxide

There are a number of nitrogen oxides (No_x, N_2O, NO, N_2O_3, N_2O_4, N_2O_5, and NO_3). Of these, nitrogen dioxide (i.e., NO_2) is the only one that is an important air pollutant. It is a water-soluble, red-brown gas with a pungent, acidic odor.

Nitrogen dioxide is produced during the combustion of fuels (gas appliances, kerosene heaters, and wood-burning stoves), tobacco smoking, and certain chemical processes. It is generally agreed that indoor pollution of NO_2 is caused primarily by cooking with gas and by cigarette smoking. A number of studies reveal the concentration of NO_2 indoors. It was found that NO_2 averages 25–400 percent more in homes with gas appliances than in electric homes (Spengler et al., 1979). In a study of houses in Chicago, concentrations of NO_2 were 40 parts per billion (ppb) in electric kitchens and 74 ppb for kitchens with gas stoves. In Columbus, Ohio, levels were 54 ppb in gas kitchens and 18 ppb in electric homes, with outdoor levels at 29 ppb (Elkins et al., 1974).

All studies have shown that indoor concentrations of NO_2 are lower than in the immediate outdoors when there are no indoor sources, due to the reactivity of NO_2 with indoor surfaces. When there are indoor sources, NO_2 levels are always higher than those outdoors.

In the environment, NO_2 is found only as a gas, and it enters through the human body through inhalation. Studies have shown NO_2 to be a deep lung irritant that can cause biochemical alteration and histological lung damage.

Studies have also revealed that 80–90 percent of NO_2 in the atmosphere is absorbed in the lungs upon inhalation. Susceptible subjects are affected by concentrations of NO_2 as low as 0.1 ppm. Substantial changes occur in pulmonary function in normal, healthy adults at or above concentrations of 2 ppm. Asthmatic persons with pulmonary problems may be affected when their NO_2 concentrations are even below 0.5 ppm.

NO_2 increases bronchial reactivity when measured by pharmacological bronchoconstrictor agents in normal and asthmatic persons. More studies are required to provide an accurate evaluation of the health consequences of the increased responsiveness to bronchoconstrictors. Evidence suggests that high concentrations or continued low levels of nitrogen dioxide increase the risk of respiratory infections (EPA and Consumer Product Safety Commission, 1995). Some studies have shown that children may have more colds and flu when exposed to nitrogen dioxide (EPA and Consumer Product Safety Commission, 1993).

Sulfur Dioxide

Sulfur dioxide is a colorless gas with a pungent odor (National Research Council, 1981). Indoor concentrations are almost always lower than outdoor concentrations due to the chemical reactivity of SO_2 with indoor surfaces and the presence of neutralizing agents such as ammonia, provided by human and animal functions.

There are two sources for indoor SO_2. The first is the infiltration from the outside ambient environment. A more important source for indoor SO_2 concentrations is from the burning of fuels that contain sulfur. For example, the sulfur contents of some bituminous coals range from 0.3 percent to more 5.0 percent and for some crude oils from 0.5 percent to 2.5 percent. The indoor concentration of SO_2 will depend upon the ventilation of the building. Table 5 provides the range of emission rates (in milligrams per hour [mg/h]) of SO_2 for kerosene space heaters and natural gas appliances.

It has long been recognized that SO_2 is a hazardous pollutant, and there have been attempts to control the emission of flue gases from stacks. As early as 1947, Los Angeles County limited SO_2 stack gases to 0.2 percent of total volume. The Environmental Protection Agency relates emission to a specified ground level concentration downwind from the source. Texas regulations set allowable concentrations according to density of population. Some governments have abandoned standards in favor of placing restrictions on the sulfur content of fuels. In the past 20 years, the SO_2 levels in cities have dropped 10–20 percent of former values. If the current SO_2 air quality standards are monitored, the future health effects of outdoor SO_2 will be minimal.

Inhalation is the only way that a person can be exposed to the effects of SO_2. At an atmospheric level of 0.03 ppm, SO_2 can be detected by most people. At a concentration above 0.12 ppm, SO_2 causes breathing distress that is noticed within a day or so. The SO_2 is absorbed in the mucous membranes of the nose and upper respiratory level. The amount absorbed depends upon the concentration. Some of the SO_2 is neutralized before entering the respiratory system due to the normal existence of endogenous

TABLE 5
Sulfur Dioxide Emission Rates from Heating Appliances

Appliance	Emission Rate (mg/h)
Kerosene space heaters	
Radiant	31–109
Convective	34–94
Gas appliances	
Range	1.29–1.66
Oven	0.67–1.09

Source: U.S. Department of Energy. *Indoor Air Quality: Environmental Information Handbook.* DOE/EV/10450-1, Washington, DC: U.S. Department of Energy, 1985.

ammonia in the human system. From the respiratory tract, the remaining SO_2 enters the bloodstream. Elimination ultimately occurs through the urine after the SO_2 is transformed to a sulphate in the liver. The sensitivity of exposure to SO_2 for short periods varies considerably for individuals. Asthmatics are strongly affected.

Studies have also shown that exposure to low concentrations of SO_2 can increase the intensity of chronic bronchitis, especially in cigarette smokers. Pulmonary diseases are also increased when there is a combined exposure to sulfur dioxide and particulates.

When SO_2 comes into contact with water, it is altered to sulphuric acid and other sulfates. These chemicals have been found in both the sensory and respiratory areas of humans. Respiratory effects from exposure to sulfuric acid have been reported to include increased respiratory rates and decreased maximal inspiratory and expiratory flow rates.

Bioaerosols

Bioaerosols are a fundamental part of the total ecosystem. They include bacteria, viruses, fungi, amoebae, algae, mites, pollen, protozoa, insect parts and feces, and arthropods. Bioaerosols are sometimes designated as microorganisms. They are found both outdoors and indoors. It is essentially impossible to control their concentration outdoors, but indoors their growth and concentration can be controlled. This is critical, for it has long been recognized that bioaerosols are responsible for various human diseases.

In order to thrive, bioaerosols require a reservoir (for storage), an amplifier (for reproduction), and a means of dispersal. In the indoor environment, bioaerosols develop in warm, moist areas such as carpets, walls, and stagnant water in air conditioners, dehumidifiers, cooling towers, and drip pans. Some bioaerosols, such as viruses and some bacteria and fungi, need a host for survival. Human bodies provide a desirable environment for growth of viruses and bacteria indoors. Areas within ductworks and air ventilation fixtures such as water pans, humidifiers, and other water sources may be sites of microbial growth.

Bioaerosols, although found in both outdoor and indoor environments, normally have their greatest concentration indoors.

Outside bioaerosols may be carried into buildings by air movement, but people, pets, and insects are also important carriers of the microorganisms.

There are many symptoms that reveal the presence of bioaerosols. These include sneezing, watery eyes, coughing, shortness of breath, dizziness, lethargy, fever, and digestive problems. Each of the bioaerosols prompt different reactions. Hypersensitivity diseases such as asthma, humidifier fever, or hypersensitivity pneumonitis are caused by immunological sensitization to bioaerosols. Extended exposure to allergens weakens the immune system, prompting reaction to lower concentrations. Bioaerosols may be the cause of a number of infectious diseases such as hypersensitive pneumonitis and Legionnaires' disease.

Bacteria

Although bacteria have a difficult time surviving for an extended period in a warm, dry atmosphere, they persist longer when the temperature is cool and the atmosphere is damp. There are a number of sources for bacteria, such as the respiratory tract (especially during colds), air conditioners, humidifying units, cool mist vaporizers, and nebulizers.

Evaporative humidifiers, although frequently contaminated with bacteria, are less likely to produce heavily contaminated aerosols. However, disinfection of any humidifying unit is difficult to achieve and, when successful, is effective only temporarily.

Other potential sources of bacteria are flush toilets and icemakers. Bacteria can also grow in damp wood and paneling and become airborne when disturbed. Many factories processing organic materials are sources of dense bacterial aerosols.

Because of health consequences to humans, even low levels of bacterial aerosols justify serious attention. Bacterial concentrations are primarily due to the types of activity undertaken in particular places. For example, operating rooms in hospitals present a particularly sensitive environment where even extremely low levels of bacteria are unacceptable.

The atmospheric concentrations of bacteria have been studied for more than a century. These efforts have revealed that there are wide differences in numbers of airborne bacteria in indoor environments depending upon temperature and humidity conditions. In poorly ventilated homes, the number can exceed 20,000 microbes/m^3. The lowest number can be near zero. The indoor number of bacteria is usually determined by the race between death and new growth.

During the last century, there was a recognition that bacteria could be present in indoor air, providing a constant threat of pulmonary tuberculosis. Although this threat has diminished due to the use of modern medicines, several recent epidemics of airborne infections in office buildings with low ventilation rates and in homes with poorly maintained air conditioner filters have shown that the presence of atmospheric bacteria remains a constant threat to health.

Viruses

Viruses are responsible for many diseases transmitted through the air. Through respiratory transmission the viruses, with sizes ranging from about 0.3 millimeters (mm) to 0.02 mm, are transferred from person to person. Common diseases carried by this means are the common cold, influenza, rubella (measles), varicella (chicken pox), and variola (smallpox).

A virus is able to survive only in living tissue. Since it does not produce its own enzyme to provide nutrition, its food source is living tissue. In the process of securing food from the living cell, viruses change the cell's normal function to that of a virus production factory, ultimately destroying the living cell while allowing more viruses into the host body. In this process the person with the virus contaminates the atmosphere by releasing the growing number of viruses in the body. For example, an entire room of schoolchildren can be contaminated with a disease such as mumps or measles within a very brief period by a single person with a contagious disease that is spread through viruses.

Fungus

Fungus has become important because of its universal presence indoors. Although there are about 180,000 known species of fungi, most of the fungi are saprophytic, utilizing dead organic material for food, but some fungi require specific types of materials for existence, such as dung and wood-rotting materials. Fungi are produced over a wide range of temperatures. Some grow at temperatures as low as 36°–40°F, whereas others will thrive at temperatures above 95°F. The prevalence of fungi indoors is more directly related to building moisture levels. Since mold spores require a high relative humidity to germinate, a minimal level of relative humidity of 70 percent (Morey et al., 1984) is required for rapid growth. Damp, nonliving organic material is quickly colonized. Condensation is the principal source of the moisture required for growth. Besides surface condensation, interstitial condensation within porous building materials such as brick, plaster, and cement may provide a reservoir of water for fungi growth. Interior dampness problems are often related to inferior construction methods. Because the water content varies greatly from season to season, fungi growth also varies greatly.

The concentration of fungi can thus vary greatly indoors. The number of colony forming units (CFU) may range from ten to more than 20,000 and may even reach higher than 400,000. However, spores and other viable colony units may account for only 1–2 percent of the total number of fungal particles. Fungi in house dust is also an important source of airborne spores indoors. Outdoor levels of fungi are usually higher than indoor levels during summer months. Indoor levels usually range from 20 to 300 percent of those outdoors and average 40–50 percent.

Studies show that fungi release disease-causing toxins (EPA and Consumer Product Safety Commission, 1995). It has been shown that fungi play a causative role in individual cases of respiratory allergies such as asthma and rhinitis. Fungi spores may also colonize the air passages of humans. Elimination of fungal growth is necessary for spore-sensitive persons. Holmberg (1987) found that airborne levels greater than 50 colony forming units per cubic meter (CFU/m^3) of certain fungi provided a significant risk factor for irritation of the eyes and respiratory tracts. In addition, indoor fungi are associated with increased fatigue, sensitization, skin irritation, and increased frequency of common colds. There is increasing evidence that respiratory diseases in children may result from mold exposure in the homes. Other important fungal infections known to be transported by air include histoplasmosis, coccidiomycosis, blastomycosis, and cryptococcosis. All cause systemic disease in unselected exposed populations.

Arthropods

Arthropods (or insects, that is, phylums of invertebrates with segmented bodies and jointed limbs) are found in both outdoor and indoor environments. Because they are toxic they have the potential to contaminate air, water, and food. Although millipedes, centipedes, sowbugs, and arachnids (including spiders) are found in almost every home, their influence on health has received little attention. One study indicates that respiratory allergy may be due to daphnia and that other health effects may be due to pheromones and similar volatile agents released from arthropods.

Protozoa

A number of studies have reported the existence of protozoa (a single-celled microscopic organism) in indoor environments. They thrive particularly in stagnant water in humidifier reservoirs and improperly drained condensation trays. Aerosolization of cells occurs directly from the contaminated water. As a result, potent allergens may develop, which if inhaled by susceptible people give rise to a hypersensitivity response resulting in pneumonitis, humidifier fever, asthma, or allergen rhinitis.

Mites

A mite is a very tiny organism that is transported by dust particles in the atmosphere. House dust mites are found everywhere in the world. They are most often associated with bedclothes, mattresses, carpet, stuffed toys, and heavily used upholstered furniture. In addition, buildings used to store or process agricultural products often get large mite populations. The natural food source for house dust mites seems to be skin scales, primarily from humans, or fungi growing on the skin. However, there are many other food sources.

The number of mites in dust varies greatly, from 10 to more that 1,000 mites per gram (g) of house dust. House dust mites, to flourish, require particular conditions of temperature and humidity. The optimal climate conditions appear to be a temperature of 77°F (25°C) and a relative humidity of 70–80 percent. The high humidity appears to be the most critical of the two factors. Some studies have reported seasonal and geographic variations in the number of mites in house dust. Consequently, the highest number of mites are found in summer when high temperatures and humidity prevail. In most instances, mites are best reduced in number by controlling dust in buildings.

Because of the great variations in number of mites from one location to another, the amount of mite allergens varies greatly among locations. However, studies of the levels of mite allergens in houses have not correlated well with the number of mites in house dust. For example, the number of dead mites may remain high until January, even though living mite numbers decline greatly after September as cooler, drier weather occurs. It must be recognized that dead mite bodies still possess allergenic properties.

Because the importance of house dust mites in human health is tied to their allergenic properties, dust extracts give relevant, positive reactions in skin tests to asthmatics. Allergy to mite allergens is common in the allergic population. Studies have revealed that 45–85 percent of asthmatic persons are affected by the presence of mites. For persons with noticeable allergy from house dust mites, exposure to 500 mites/g dust is considered a level to provoke acute asthma attacks.

Allergens

Most animals, such as dogs and cats, produce allergens that can be inhaled by humans. The allergens produced by pets are associated with dander, hair, saliva, and urine (Lowenstein et al., 1986). Allergens found in dander and hair become airborne dust particles that are 1–10 mm in diameter. Of the possible allergens, only a few have been identified. Of these, the major cat allergen (Feldi) has been characterized and standardized.

The concentration of animal allergens in buildings varies greatly. In a study by Lowenstein (1986) in houses where one cat was present, concentration varied from 250 to 1,140 particles per m^3 of air. In house dust samples in houses with dogs, concentrations of 1,100 to 525,000 particles per m^3 have been found (Wood et al., 1988).

Because allergens are extremely small, concentrations build up and remain suspended in air for a long time. This may account for the rapid onset of asthma or rhinitis in patients allergic to cats upon entering a house. Studies have suggested that hypersensitivity to cat allergens in the general population reveals a frequency of positive skin reactions of 10–15 percent (Hader et al., 1990). Adverse reactions to dog dander have been found to be species specific.

In order for us to fully understand the health effects on humans, many more allergens of the animal world must be characterized and standardized. In 1991, the first dog hair and dander-specific allergen (Can fl) was isolated (Schou et al., 1991). It may now be possible to use Can fl as a marker allergen for future studies of the relationship between exposure and sensitivity of humans.

Algae

Algae cells are abundant in the outdoor atmosphere. The cells are best developed in a water habitat but are also found in soils and damp natural surfaces. They are carried by air movement into indoor environments and are often present in house dust. It has been well documented that human air passages are sensitive to algae but their total impact remains unclear.

Insect Emanations

Many insects have emanations that are strong sensitizers capable of causing respiratory allergy. Cockroaches have received special attention because of their abundance and their ability to cause respiratory symptoms. The roach's fecal pellets appear to provide antigens (substances that create antibodies when introduced into the blood) for the insects eating them; on dissolution, the pellets may contribute importantly to air contamination. This effect is most noticeable in areas of poor sanitation. Other indoor insects such as houseflies, bedbugs, and carpet beetles are known to produce respiratory allergies in exposed persons.

Pollen

Most pollens are produced by plants outdoors. However in homes, the pollens of indoor plants like cyclamen and impatiens have been found to reach levels of hundreds of grains per cubic meter. The indoor pollen is increased by wind movement of outdoor pollens. Once inside the house, the microscopic pollen grains remain for extended periods of time. Indoor pollen is best controlled by ventilation. The effect of pollen on health has not been studied.

Physical Pollutants

There is a wide variety of physical pollutants with different characteristics, varying from radioactive radon to single dust particles. Many, such as lead, are highly toxic. Others, such as asbestos, are contaminants, because their tiny silica minerals destroy the function of the lungs.

Radon

Origin of Radon Radon is a radioactive noble (inert) gas, which exists in several isotopic forms. Radon originates from the decay chain of uranium^{-238} and thorium^{-232}. In the decay of uranium^{-238},

$radon^{222}$ is produced; from thorium, $radon^{220}$ is the product. In the decay chain, radon is the first and only gaseous and inert element produced. Of the two radon elements, $radon^{222}$ is most important, for it has a lifetime of 3.8 days compared to the shorter lifetime of $radon^{220}$ (55 seconds). $Radon^{222}$ decays by the emission of alpha (a), beta (b), and gamma (g) rays. In the decay of $radon^{220}$ a number of short-level nuclides (radon daughters) are produced, exhibiting a half-life of about 30 minutes.

The radon content of regions varies considerably depending upon the uranium content of the bedrock and soil. The Environmental Protection Agency has estimated that the average soil contains about 1 ppm of uranium. Phosphate rock contains 5–12 ppm, granite, as much as 50 ppm.

Indoor Concentration Outdoor air concentrations of radon are small (and so the possibility of outdoor radon being a health hazard is extremely small), but indoor concentrations are typically two to ten or more times higher. Indoor concentrations depend upon the rate of entry from bedrock sources and the rate of removal by ventilation (Miller and Miller, 1990).

It has been determined that the entrance of radon into a building depends upon the radon production rate in the bedrock at the soil, meteorological factors, soil permeability, the type of building substructure, and the so-called stack effect (when radon moves from higher pressure inside the earth to the lower pressure of the building). The stack effect is a major factor in radon infiltration into buildings (EPA, 1994).

Since 1980 many studies have been conducted on indoor radon levels (Martinowski, 1992). These surveys reveal that the indoor radon levels in buildings are normally in the range of 10–140 becquerels per cubic meter (bq/m^3). However, it is impossible to estimate the maximum indoor radon concentration in a particular place. Concentrations as high as 100,000 bq/m^3 have been detected in individual dwellings. The Commission of European Communities has indicated that any level of radon above 400 bq/m^3 can be detrimental to health. As a result, it has been estimated that excess radon is found in 0.5–3.0 percent of European dwellings.

Detection of Radon Not until 1984 was it recognized that health hazards from radon were extremely widely distributed over the earth. This was triggered by the now famous incident at Boyerstown near Reading, Pennsylvania. On a December morning in 1984, a worker at the Limerick nuclear power plant set off an alarm when he went to work. It was soon proved that the radiation did not come from the plant but from his home. The home was tested for radiation, and 3,200 picocuries (one-trillionth of a curie, a curie being equal to 37 billion [37×10^9] disintegration of radioactivity per second) were discovered in the living room. Up to this time, no one realized that radon could be a danger to anyone except uranium miners and people who lived near uranium mines. A totally new concept came into existence—that radon could be constantly released from the soil, infiltrate houses, and accumulate in life-threatening concentrations.

Due to the very high picocurie readings at Boyerstown, testing for radon contamination spread across the nation. Some of the highest levels of radon were found in Pennsylvania, New Jersey, Wisconsin, Minnesota, North Dakota, and Colorado. The Environmental Protection Agency and the Public Health Service estimate that radon poses a health problem in 8 million U.S. homes. The EPA further estimates that 5,000–20,000 lung cancer deaths annually are due to radon exposure.

Health Hazard On the basis of such studies, radon has proved to be a human carcinogen. The principal health risk comes from the inhalation of radon and its daughters. The radiation dose to living tissue is dominated by the alpha particles emitted by the deposited products. From a health perspective, the short-lived daughters, polonium^{-214}–polonium^{-218}, are of highest significance. Thus, the contribution to lung dose due directly to radon gas is small. Radon acts mainly as a "carrier" of radon daughters, which produce the radioactive effect. The radon daughters are most dangerous because of their short life. They decay directly on the membranes of the lungs, which are particularly sensitive to the burning effect of radiation. Other environmental factors such as exposure to mineral dust and chemical aerosols may increase the damage to surface cells, thus increasing sensitivity to radiation.

Because radon and radon daughters can cause cancer, they are now recognized as a major health concern. Acute exposure can damage other organisms and affect gastrointestinal functions and the nervous system, resulting in internal bleeding, cardiac distress, and sometimes death. Exposure to low doses over extended periods may produce delayed effects such as cell transformation. Mutations can be deleterious in many ways, such as reducing resistance to diseases (including cancer).

Asbestos

Uses of Asbestos *Asbestos* is a generic term that applies to a group of impure hydrated silicate minerals. During the past century, the use of asbestos has increased from a few pounds to a peak of 4 million tons worldwide. Asbestos is used to insulate many products, from hair dryers to steel trusses, against fire. It is also used in the manufacturing of building tiles, roofing shingles, brake linings for automobiles, and fire-protection garments. Asbestos was a common insulation in school buildings.

Natural Sources Natural sources are important because asbestos minerals are widely distributed throughout the earth's crust. The air above natural deposits can reach 48 fibers/liter (l). Water from asbestos deposits may also contain asbestos fibers. In particular, chrysolite is present in most serpentine formations. Very little is known, however, about the amount emitted from natural sources.

Asbestos Fiber Emissions Man-made emissions of asbestos originate from a number of activities. Asbestos fibers are emitted into the atmosphere from mining, milling, and the manufacturing of products. Construction activities are also a source of emissions, as are transport and use of asbestos products. Asbestos is frequently exposed on surfaces in buildings, and over time these surfaces may become damaged, causing asbestos fibers to fray and then be emitted into the atmosphere. This occurs especially in school buildings where children damage the walls. Asbestos may also be emitted into the atmosphere when waste asbestos is disposed of.

The identification of asbestos fibers in the environment is extremely complex. It requires an integrated procedure including microchemical analysis of single fibers, measurement of fiber length and diameter, and counts of fibers in a given area. Electron microscopy is the only method that can currently detect and identify asbestos fibers among the very wide range of other fibrous and nonfibrous particles in the atmosphere. To positively identify asbestos fiber, an X-ray diffraction analyzer must be attached to the electron microscope, a very costly procedure. As a consequence, the number of studies measuring atmospheric asbestos remains limited (Maroni, Seifert, and Lindvall, 1995).

The concentration of asbestos fibers varies greatly from place to place. Studies conducted in different environmental settings reveal:

Rural areas (remote from asbestos emission sources): Below 100 F/m^3 (National Research Council, 1984).

Urban areas: Fiber levels vary from below 100 to more than 1,000 fibers per cubic meter (F/m^3)in buildings (Commins, 1985).

Indoor air: In buildings without specific asbestos sources, concentrations are generally below 1,000 F/m^3 (Commins, 1985). For buildings with friable asbestos, concentrations vary greatly. The concentration of asbestos fibers is usually less than 1,000 F/m^3 but can reach 10,000 F/m^3 (Commins, 1985).

Special environment: At 1,000 feet downwind from an asbestos-cement plant, concentrations are 2,200 F/m^3; at 2,200 feet, 800 F/m^3; at 3,000 feet, 600 F/m^3. Street crossings with heavy traffic have concentrations averaging 900 F/m^3 (Sawyer, 1979). On an expressway, concentrations of up to 3,300 F/m^3 can occur.

Workplace air: Traditional levels of asbestos fibers in workplaces were orders of magnitude higher than those found in the environment, with values from 10^5 F/m^3 to more than 10^8 F/m^3 (Commins, 1985), but legislation controlling the use of asbestos has reduced the levels to below 2 x 10^8 F/m^3 in most countries.

Health Hazard The health danger of asbestos fibers depends largely upon the number of fibers in the atmosphere and the number of years of exposure. The most serious health effect from exposure to asbestos fibers is lung cancer. Studies have shown that when smokers are exposed to asbestos fibers, the risk of cancer is greatly increased. In addition, the risk of lung cancer appears to increase for workers in the mining and milling of asbestos as well as those in the production of asbestos cement and textiles. The greatest danger, however, is to occupants of buildings in which asbestos has been damaged and fibers released into the atmosphere.

The most serious of the asbestos-generated cancers is mesothelioma. It attacks the tissues of mesothelial origin. Although a rare type of cancer, it is essentially always related to occupational asbestos exposure. The time elapsing between first exposure and the clinical appearance of a tumor ranges from 20 to 50 years. It creates few symptoms for years, but mesothelioma, once diagnosed, is incurable (National Research Council, 1984).

Heavy Metals

Metals found in the indoor environment in significant quantities include lead, mercury, arsenic, and cadmium. The health hazards associated with trace levels of heavy metals is well documented by the Agency for Toxic Substances and Disease Registry.

Lead

Sources of Lead Among various metals, lead was one of the first to be used by humans, who have smelted lead for more than 4,000 years. Lead initially was used for many household items such as plates and cups. Consequently, many unsuspecting people of the ancient world were poisoned by lead in their food and drink.

In the twentieth century, lead exposure results from a number of sources. Excessive lead in the body may come from food and water and the breathing of lead particles in the atmosphere. According to the EPA, exposure to lead-based paint "is the most significant source of lead exposure in the [United States] today" (EPA and Consumer Product Safety Commission, 1995). Lead was once a major ingredient in the production of paint and was used in homes built prior to 1960. However, some homes built as recently as 1978 may also contain lead-based paint (EPA and Consumer Product Safety Commission, 1995). As the lead oxidizes, it has a sweet taste. As a result, children find the taste of paint chips appealing and become poisoned when they swallow them. This is primarily a problem in older homes and schools.

Federal Legislation The problem of indoor lead poisoning was recognized by the U.S. Congress in 1971, when it passed the Lead-Based Paint Poisoning Prevention Act. This law provided federal assistance to develop and implement local programs to eliminate the causes of lead-based paint poisoning (see Chapter 2). The law has been amended several times since it was passed, but the control of lead-based paint indoors has proven to be a very difficult problem. According to the Agency for Toxic Substances and Disease Registry, about one in six children in the United States have high levels of lead in their blood (EPA Report, 1992).

Control of Lead To eliminate the danger of lead poisoning in buildings, any lead-based paint must be removed. The traditional methods for removal are scraping and sanding and require special equipment, trained personnel, and proper procedures. But this process generates a significant amount of particles, requiring thorough postremoval cleanup of floors and other surfaces. When inexperienced people, such as homeowners, use this process to remove lead paint from walls, no systematic or standard procedure has been developed to control the highly dangerous debris. In recent years, an open flame or heat gun has been used to remove paint, but this procedure has been found to be unsatisfactory. When a painted surface is heated, the lead vaporizes from the paint and creates an additional health hazard. Even when lead paint is removed by sanding or by flame, a portion remains. The lead content of house dust after using these methods was found to be at preabatement levels 6–12 months following removal. Only complete paint removal provides a lead-free environment. Homes must be tested after removal to ensure that all residual lead has been removed. Guidelines have been established by the U.S. Centers for Disease Control in Atlanta to guarantee a lead-free environment. Consequently, the use of lead paint has been discontinued. However, the lead paint in old buildings will remain a problem for decades.

Other Lead Sources There are other sources of lead in the atmosphere. Until the 1970s, tetraethyl lead was used as an antiknock gasoline additive. Upon ignition, the lead dispersed with other combustion gases into the atmosphere and entered houses near heavily traveled roads. Consequently, traces were found in the blood of people in these locations. In 1974 the Environmental Protection Agency completed a thorough evaluation and added lead to the list of primary ambient pollutants (EPA, 1977). Since 1974 national legislation has required that all new motor vehicles operate on unleaded gasoline. Leaded gasoline has essentially disappeared. Lead levels in the atmosphere have thus diminished greatly, and health hazards caused by lead have disappeared. Until the early 1970s printers were exposed to high lead levels due to lead in ink. A new newsprint was developed that contains only about 0.53 ppm lead in black and white print and 0.72 ppm in colored print. The lead resulted in industrial pollution, and when the paper was burned in the home lead was released into the atmosphere. There are also several other sources for indoor lead pollution. Lead is emitted into the atmosphere from the smelting of lead ores, combustion of fossil fuels, and manufacturing of lead products.

The human body contains some lead. On a normal day, the human intake is approximately 0.45 mg, of which about one-third is from the air and two-thirds from water and food. A study of body retention of daily lead intakes revealed that about one-half remained within the body (Rabinowitz et al., 1975). The lead remaining in the body varies from 80 mg to 200 mg depending upon local pollution.

Health Hazard The effect of lead on health depends upon the amount and extent of exposure. The principal target organs or organ systems include the blood, the brain, and the nervous system, the reproductive system, and the kidneys. Symptoms from acute exposure may include colic, shock, severe anemia, acute nervousness, brain damage, and kidney dysfunction. There is also evidence that lead poisoning leads to sterility, spontaneous abortions, stillbirths, and neonatal problems. Because lead appears to inhibit development of enzymes in the pathway of hemoglobin biosynthesis, hematological changes (effects on the blood) are the first detectable manifestations of low-level chronic lead exposure. This decreased hemoglobin production is in part the cause of the anemia associated with chronic lead poisoning. Severe lead poisoning may lead to death (EPA and Consumer Product Safety Commission, 1995).

Lead poisoning affects not only children but women of child-bearing age, who are at special risk. The deleterious effects on reproduction have proven serious. As lead reaches the placenta, high blood lead levels in the mother may enter the fetus, potentially resulting in mental retardation. This problem is of special concern in the industries using lead, as women enter the workforce of previously male-dominated occupations.

Mercury

Mercury has been used for centuries to treat certain types of illnesses. Its use as a medicine was based on the assumption that diseased tissues were less resistant to poison than healthy ones. Consequently, it was believed that infected tissues would be killed before the entire organism would die. There was also the hope that mercury would stimulate the formation of generally effective defenses against further diseases. As late as 1940, mercury was the only effective treatment for syphilis.

Mercury in the indoor atmosphere comes from a variety of sources. The main sources are wastes containing mercury residue. Until the passage of the Clean Water Act in 1972, paper and chlorine manufacturers discharged mercury wastes directly into streams and lakes. Although this practice is no longer permitted, the environmental damage of mercury residue in streams and lakes will persist for many years, and there remains a toxic danger to anyone drinking tainted water. About 1 million pounds of mercury are discharged into the atmosphere each year from fuel combustion. Atmospheric mercury levels currently range from 0.6 mg/m^3 over the oceans to 1,200 mg/m^3 above mercury deposits. Urban areas contain about 5 mg/m^3 (National Research Council, 1978).

The major nonoccupational source of mercury comes from the mercury amalgam prepared in dental offices. Measured air concentrations in such offices range from 0.01 mg/m^3 to 0.18 mg/m^3. Scientific laboratories that use mercury may also have high levels. The greatest danger of excessively high levels occurs after spills in these areas. Mercury is so fluid that tiny particles can become trapped in the floors of rooms for many years. The toxic effect of mercury can be neutralized by using elemental sulfur, which converts the metal into a more stable sulfide.

Mercury normally enters the body through the consumption of water and food. Once it enters the body it is moved by the circulatory system and accumulates in the liver, kidneys, and central nervous system. Mercury is removed from the body in urine, feces, and bile. A high level of mercury can cause chronic poisoning. The past effects of mercury poisoning are well known from the slogan "mad as a hatter" (mercury was once used to make hats). Not only mental deterioration but also death can result from poisoning.

Cadmium

Cadmium is highly toxic to humans. A major indoor source of cadmium is tobacco smoke. Approximately 1 mg of cadmium exists in a pack of cigarettes. Cadmium also enters the body through foods. In such industrial processes as welding and soldering activities, cadmium fumes can be inhaled. Cadmium is also found in the atmosphere. The body's retention of cadmium from the atmosphere is much higher than that received from foods.

Cadmium causes acute and chronic illnesses from deposits in the kidneys and liver. It can severely damage capillaries in the kidneys and will interact with nutrients in the liver. Cadmium is also among the environmental contaminants linked to renal disease.

Particulates

Physical particles in the atmosphere include dust and man-made mineral fibers.

Dust

House dust is found throughout the world under all atmospheric conditions. Physical dust particles, originating from such sources as rock, wood, grains, and ores, are produced by handling, crushing, grinding, impact, detonation, and decrepitation; the sources of indoor dust thus vary tremendously. Buildings located on unimproved roads are likely to have large quantities of rock dust. In dry regions, such as deserts, windstorms carry dust hundreds of miles. Special sources of outdoor dust that gradually enter buildings include volcanic eruptions and atomic explosions. Dust particles, depending upon their origin, vary greatly in size, ranging from 0.1 mm to the size of grains of sand. It is normal for indoor dust levels to exceed outdoor levels because the tiny dust particles are difficult to remove by any type of ventilation filters.

However, dust particles between 0.1 mm and 1.5 mm are most important to human health, for these sizes penetrate and deposit permanently in the respiratory tract. Dust is a major cause of asthma and rhinitis. If there is long exposure to dust, breathing is affected. Dust may also cause coughing, and susceptibility to common colds is increased. Ultimately the dust particles may lodge on the lymph nodes, where they become embedded for life, creating a chronic health problem. Depending upon the chemical components of the dust, it may dissolve in the bloodstream, where it can remain for years.

Man-Made Mineral Fibers

Man-made mineral fibers are fibers manufactured from glass, natural rock, or other minerals. They are composed of the oxides of earth materials and are classified as to their source materials such as glass wool, rock wool, or slag wool. Categories and materials are named with reference to the raw material process, structure, and use.

Man-made mineral fibers have a wide variety of uses. Of the more than 6 million tons of mineral fibers produced annually, fibrous glass accounts for about 80 percent. In recent years, fibrous glass has been widely used as a substitute for asbestos in acoustic and thermal insulation (which accounts for about 85 percent of its use). About 10–15 percent of fibrous glass is used in textile production for the reinforcement of resinous materials and draperies. About 1 percent of all glass fiber is used in specialty applications such as high-efficiency filter paper and insulation for aircraft (World Health Organization, 1988).

The health effects of man-made mineral fiber were recognized as early as the beginning of the twentieth century. Fibrous glass and rock wool (mainly larger than 4.5–5.0 mm in diameter) cause physical irritation of the skin characterized by itching erythema. The irritant dermatitis reaction may be complicated by urticarial and eczematous reactions (transient skin disorders and inflammatory diseases of the skin, respectively) that appear as an allergic response. Some people have an allergic reaction to resins used in man-made mineral fiber. A few recent studies have also reported eye irritation to mineral fibers. One study reports that man-made mineral fibers may cause nonmalignant respiratory disease, and in high occupational exposure carcinogenic effects have been observed (World Health Organization, 1988).

Volatile Organic Compounds

Volatile organic compounds (VOCs) are found in all indoor environments. The number of identified VOCs has risen steadily in recent years, from 250 to more than 900 in 1989 (EPA, 1989) to well more than 1,000 currently. Volatile organic compounds have a wide range of physical and chemical characteristics. Of these, two are particularly important: water solubility (i.e., hydrophillic character) and whether the VOC is neutral, basic, or acidic. Water solubility is important because it determines the retention of solid materials.

Volatile organic compounds are released into the indoor environment by a wide variety of materials such as consumer products, house furnishings, pesticides, and fuel (see Table 6).

Concentrations vary tremendously depending upon the nature of the product. In many areas, outdoor contamination contributes to the levels of VOCs indoors. More than 250 organic chemicals have been identified in indoor air at levels greater than 1 ppb, with VOC concentrations ranging from a few parts per billion to a few parts per million.

It is now known that exposure to many VOCs causes acute and chronic health effects. Although there are differences in individual responses, the health effects may include fatigue, dizziness, weakness, skin irritation, blurred vision, irritability, and other symptoms (see Table 7).

A number of studies have revealed that some VOCs are human carcinogens (benzene) or animal carcinogens (carbon tetrachloride, chloroform, trichloroethane, tetrachlorveth, and 1,4-dichlorobenzene). Other VOCs, such as octane, decane, and n-decone, are possible cocarcinogens. Some VOCs (1,1,1-trichloroethane, styrene, and a-pinene) are known mutagens (EPA, 1989).

TABLE 6
Sources of Volatile Organic Compounds in Indoor Air

Sources	Contaminants
Consumer and commercial products	
Cleaners and waxes	aliphatic hydrocarbons
liquid floor wax,	aromatic hydrocarbons (toluene, xylenes)
aerosol furniture wax,	halogenated hydrocarbons (tetrachlorethane, methylene chloride, 1,1,1-trichloroethane, 1,4-dichlorobenzene, alcohols)
aerosol window cleaner,	ketones (acetone, methyl ethyl ketone)
and others	aldehydes (formaldehyde)
	esters (alkyl ethoxylate)
	ethers (glycol ethers)
	terpenes (limonene, a-pinene)
Paint and associated items	
paint, varnishes,	aromatic hydrocarbons (toluene)
paint thinners,	aliphatic hydrocarbons (n-hexane, n-heptane)
paintbrush cleaners,	halogenated hydrocarbons (methylene chloride, propylene dichloride)
paint removers, and other items	alcohols
	ketones (methyl ethyl ketone, methyl isobutyl ketone)
	esters (ethyl acetate)
	ethers (methyl ether, ethyl ether, butyl ether)
Adhesives	
rubber cement, plastic	aliphatic hydrocarbons (hexane, heptane)
model glue, floor tile adhesive,	aromatic hydrocarbons
all-purpose adhesives	halogenated hydrocarbons
	alcohols
	organic nitrogen compounds (amines)
	ketones (acetone, methyl ethyl ketone)
	esters (vinyl acetate)
	ethers
Cosmetic/personal care products	
perfume, deodorants, shampoo,	alcohols (propylene glycol, ethyl alcohol, isopropyl alcohol)
soap, rubbing alcohol, hair sprays	ketones (acetone)
	aldehydes (formaldehyde, acetaldehyde)
	esters
	ethers (methyl ether, ethyl ether, butyl ether)
Automotive products	
motor oils, gasoline, automotive	alphatic hydrocarbons (kerosene, mineral spirits)
cleaners, hydraulic fluids, motor	aromatic hydrocarbons (benzene, toluene, xylenes)
vehicle exhausts	halogenated hydrocarbons (tetrachlorethane)
	alcohols (ethylene glycol, isopropyl alcohol)
	ketones (methyl ethyl ketone)
	amines (triethanolamine, isopropanolamine)
Hobby supplies	
prophatic chemicals, adhesives,	alphatic hydrocarbons (kerosene, hexane, heptane)
molding clays, wood fillers, other	halogenated hydrocarbons (methylene chloride, ethylene chloride)
items	alcohols (benzyl alchohol, ethanol, methanol, isopropyl alchohol)

(continues)

TABLE 6
(*continued*)

Sources	Contaminants
	aldehydes (formaldehydes, acetaldehyde) ketones (methyl isobutyl ketone, acetone) esters (di-[2-ethylhexyl] phtalatel) ethers (ethylene glycol ether) amines (ethylene diamine)
Heating, ventilation, and air-conditioning systems, combustion appliances furnaces, unvented heaters, gas cooking stoves, fireplaces, air conditioners	aliphatic hydrocarbons (propane, butane, isobutane)
Furnishings and textiles carpets, upholstered furniture, plastics, draperies, blankets, mattresses, and the like	aromatic hydrocarbons (styrene, brominated aromatics) halogenated hydrocarbons (vinyl chloride) aldehydes (formaldehyde) ethers esters
Building materials pressed wood products, adhesives, plastics, wall coverings	alphalic hydrocarbons (n-decane, n-dodecane) aromatic hydrocarbons (toluene, styrene, ethylbenzene) halogenated hydrocarbons (vinyl-chloride) aldehydes (formaldehyde) ethers ketones (acetone, butanone) esters (urethane, ethyl acetate)
Tobacco smoke	VOCs organic nitrogen compounds (nicotine) aldehydes (formaldehyde, acrolein) ketones
Human and biological products animal feces, pets, plants, metabolic products, pathogens	aliphatic hydrocarbons (methane) aromatic hydrocarbons (toluene) aldehydes (acetaldehyde) ketones (acetone, 2-hexanone) alcohol (3-methyl-1-butanol) pesticides
Pesticides aerosol household pesticides, roach killer, flea killer, mold and mildew inhibitors, houseplant insecticides, moth repellant, rodenticides, fungicides, others	aliphatic hydrocarbons (kerosene) aromatic hydrocarbons (xylene) halogenated hydrocarbons (chlordane, 1,4-dichlorobenzene, heptachlor, chloropyrifos, diazinon) ketones (methyl isobutyl ketone) organic sulfur/phosphorous compounds (malathion)

Source: U.S. Environmental Protection Agency, *Preliminary Indoor Air Pollution Information Assessment.* Appendix A, Report No. EPA-600/8-87/014, Washington, DC: Environmental Protection Agency, 1987, pp. 2–19.

TABLE 7
Health Effects of Selected Volatile Organic Compounds

Compound	Health Effects
Benzene	Carcinogen; repiratory tract irritant
Xylenes	Narcotic; irritant; affects heart, liver, kidney, and nervous system
Toluene	Narcotic; possible cause of anemia
Styrene	Narcotic; affects control of nervous system; probable human carcinogen
Toluene diisocyanate (TDI)	Sensitizer; probable human carcinogen
Trichloroethane	Affects central nervous system
Ethyl benzene	Severe irritation of eyes and respiratory tract; affects central nervous system
Dichloromethane	Narcotic; affects nervous system; probable human carcinogen
1.4-Dichlorobenzene	Narcotic; affects liver, kidney, and central nervous system; eye and respiratory tract irritant
Benzyl chloride	Central nervous system irritant depressant; affects liver and kidney; eye and respiratory tract irritant
2-Butanone (MEK)	Irritant; central nervous system depressant
Petroleum distillates	Affects central nervous system, liver, and kidneys
4-Phenylcyclohexene	Eye and respiratory tract irritant; central nervous system effects

Source: U.S. Environmental Protection Agency (EPA). *Introduction to Indoor Air Quality.* Report no. EPA/400/3-91/003, Washington, DC: U.S. Environmental Protection Agency, 1991.

A number of studies have indicated that VOCs play a role in sick building syndrome. Mucous membrane irritation and impaired memory in otherwise healthy subjects. Many humans exposed to n-decone in the range of 0–100 ppm experienced mucous membrane irritations, decreased tear film stability, and increased odor intensity (Kjaergard et al., 1987).

Many VOCs are potent narcotics that cause a depression in the central nervous system. Other VOCs cause an irritation of the eyes and respiratory tract. Some people are affected by multiple chemical sensitivity, which is caused by breathing VOCs and other organic compounds released by building materials and consumer products such as cosmetic soaps, perfumes, tobacco, plastics, dyes, and other products (Ashford and Miller, 1991).

Formaldehyde

Formaldehyde (HCHO) is the most common of the aldehydes. It is now recognized as possibly the single most important indoor pollutant because of its common occurrence and its strong toxicity. At normal room temperatures, it is a colorless gas with a pungent odor. It is soluble in water. Under normal atmospheric conditions, formaldehyde is quickly photooxidized in sunlight to carbon dioxide. Outdoors, it has a half-life of approximately 50 minutes during daytime. When nitrogen oxide is present, the half-life drops to about 35 minutes. Formaldehyde in the indoor atmosphere has a much longer life span.

Indoor Sources Formaldehyde is released into the indoor environment from a number of sources. Of these, wood products are among the most important sources. Urea-formaldehyde resins are commonly used to manufacture particleboard, plywood, paneling, furniture, wallboard, and ceiling panels. Formaldehyde is a highly water-soluble gas that can bind together wood and other hygroscopic (water-absorbing) surfaces. Because formaldehyde is inexpensive, it is the most common adhesive used to produce wood products.

Formaldehyde resins are also used in a number of household items. Formaldehyde resins used in the manufacture of carpets, wallpapers (especially decorative wallpapers), and other papers can emit formaldehyde into the atmosphere. The textile industry uses formaldehyde resins to produce crease-proof, crush-proof, flame-resistant, and shrink-proof fabrics. Formaldehyde is also used in binders to improve the adherence of pigments to cloth (see Table 8).

Formaldehyde resins are also commonly used in the production of insulation materials. Low-density urea-formaldehyde foam insulation (UFFI) became popular in the early 1970s. Because it was considered that emission of formaldehyde could not be controlled, and because formaldehyde was considered to be a

TABLE 8
Formaldehyde in Selected Industries

Industry	Formaldehyde Level (ppm)
Fertilizer	0.2–1.9
Dyestuffs	< 0.1–5.9
Textile	< 0.1–1.4
Resins	< 0.1–1.4
Bronze	0.12–8.0
Iron foundry	< 0.02–18.3
Treated paper	0.14–0.99
Plywood	1.0–2.5

Source: National Institute of Occupational Safety and Health. "Formaldehyde Evidence of Carcinogenicity." *NIOSH Current Intelligent Bulletin* 34 (April 15, 1981): 1+.

potential carcinogen to humans, the U.S. Consumer Product Safety Commissioner on February 22, 1982, banned the sale of UFFI. However, the U.S. Court of Appeals for the Fifth Circuit thereafter issued an opinion that reinstated UFFI sales. A number of patents have been filed to reduce the release of formaldehyde gases into the atmosphere.

Cigarette smokers (and nonsmokers indirectly) are exposed to formaldehyde. Cigarette smoke contains as much as 40 ppm of formaldehyde by volume. The individual who smokes a pack of cigarettes daily receives a daily burden of 0.38 mg of formaldehyde.

Formaldehyde emissions may also come from an oven or a single gas stove flame. Other products that might contain formaldehyde or formaldehyde resins include starch-based glues, room deodorizers, cosmetics, and some toiletries. Typical paper products treated with urea-formaldehyde resins include grocery bags, waxed paper, facial tissue, napkins, paper towels, and disposable sanitary products. The practice of permitting urea-formaldehyde-treated paper to come in contact with food has been permitted by the U.S. Food and Drug Administration since 1972.

Concentration Levels Essentially everyone is exposed to formaldehyde during the course of a day. Indoor concentrations are influenced by temperature, humidity, ventilation rate, age of the building, type of material in the building, products used, and the smoking habits of occupants.

The highest concentrations of formaldehyde are found in industrial buildings. It is estimated that about 1.6 million industrial workers are exposed to formaldehyde daily. Among all dwellings, formaldehyde exposure appears to be highest in mobile homes. In the past, some prefabricated mobile homes contained elevated formaldehyde levels due to the large amounts of high-emitting pressed-wood products used in their construction as well as their relatively small interior space (EPA and Consumer Product Safety Commission, 1995). The second largest concentration of formaldehyde is found in homes insulated with UFFI. The average concentration is 0.12 ppm in UFFI homes compared to 0.03 ppm in non-UFFI homes. A high concentration of formaldehyde is frequently found in homes that are tightly sealed for energy efficiency. The concentration in these homes may exceed 0.1 ppm. The addition of such things as curtains or furniture may also increase formaldehyde levels.

A number of studies have reported that indoor concentrations of formaldehyde have decreased in homes in the past decade due to the type of construction materials used. Counteracting this trend, however, is the one toward constructing more heavily insulated buildings. The level of indoor concentration due to these competing factors makes it difficult to predict the effect of any measures to reduce the concentration levels.

Time Variations Diurnal and seasonal indoor concentrations of formaldehyde can vary considerably depending upon the rate of ventilation, the weather, and seasonal changes. A number of studies have shown that the formaldehyde level in a home diurnally can vary as much as 50 percent, with a range between 140 mg/m^3 and 300 mg/m^3 (Moschandreas et al., 1978). These changes seem to depend upon whether or not the house is insulated, ventilation rate, and the activities of occupants. In addition there are seasonal variations in formaldehyde activities. According to the EPA and the U.S. Consumer Product Safety Commission, formaldehyde levels are highest in summer and lowest in winter. Temperature control is thus evident.

Ventilation can be an important controller of formaldehyde levels in a room. If there is a high exchange rate, the outdoor concentration of formaldehyde in a room will be inversely proportional to the rate of air change. The condition exists when the windows and doors are left open. Under less advantageous exchange rates, a nearly threefold increase in the exchange rate of air induced only about a 20-percent reduction in formaldehyde concentrations.

Health Effects Formaldehyde is a strong toxic irritant. When it reaches levels of 12–25 mg/m^3, breathing becomes difficult. Effects on pulmonary tissue and lower airways are likely at concentrations of 6–40 mg/m^3, and pneumonitis and pneumonia occur at concentrations of 60–120 mg/m^3. Occupational studies have revealed that 1–2 percent of the population that is exposed to higher concentrations of formaldehyde will develop asthma. Common symptoms of formaldehyde exposure are irritation of the eyes, nose, and throat, lachrymation, sneezing, coughing, nausea, dizziness, lethargy, irritability, disturbed sleep, and olfactory fatigue. See Table 9. Symptoms may occur suddenly, with some relief after a period before they reoccur. Children are especially sensitive to formaldehyde emissions.

Since the 1970s a number of studies have attempted to assess the potential carcinogenicity of formaldehyde in humans.

The subjects investigated used formaldehyde to preserve human tissues (embalmers, anatomists, and pathologists) and industrial workers who used formaldehyde to produce a product. Formaldehyde has proved to cause cancer in laboratory animals and may cause cancer in humans.

Other studies have been carried out on the health effects of formaldehyde. These studies have addressed such problems as the relationship between formaldehyde and miscarriage of babies, frequency of menstrual disorders, pregnancy complications and low-birthweight babies, and chromosome aberrations.

Polynuclear Aromatic Hydrocarbons

Polynuclear (or polycyclic) aromatic hydrocarbons (PAHs) consist of a large number of organic compounds with two or more benzene rings. When found indoors, they exist as vapor and are absorbed on particles. Hundreds of different PAHs are ubiquitous in the atmosphere. They are generally produced by combustion of fossil fuels and by industrial processes.

In buildings, the major sources are from outdoor infiltration and indoor combustion from cigarette smoking, gas stoves, unvented space heaters, wood-burning stoves, and fireplaces. Outdoor sources include combustion of gasoline, diesel fuels, coal, and wood products. Agricultural burnings and forest fires can be significant PAH sources. Industrial processes that produce PAHs in indoor atmospheres include petroleum refining, coke production, and steel manufacturing. PAHs are also present in food and water, which are major sources for human consumption.

Although about 500 types of PAHs have been detected in the atmosphere, most measurements have been made on benzo (a) pyrene (BaP). The relationship between the amount of BaP and other PAHs determines the PAH profile. Although PAH profiles of emissions can differ greatly, they are relatively similar in the ambient air. Due to greater controls on combustion processes during the past 30 years, the natural background level of BaP has declined greatly, reaching almost zero in some cities. Gasoline combustion in vehicles prevents outdoor PAHs from declining significantly.

Indoor concentrations of selected PAHs have been recorded in a study of 33 homes in California and Ohio (Wilson et al., 1991). These varied from 0.17 mg/m^3 to 240 mg/m^3 depending upon whether there was cigarette smoking plus gas stove operation in the homes. The greatest concentration was detected in a home where there was both smoking and gas stove operation. Indoor concentrations of PAHs are also higher than concentrations outdoors.

On the basis of experimental tests, the greatest danger to exposure of PAHs is the development of skin cancer. Chimney sweeps and tar workers who were exposed to substantial amounts of PAHs have shown evidence of skin cancer. Workers on coke ovens, coal gas, and in aluminum plants have provided evidence that inhaled PAHs caused lung cancer. It is, however, recognized that not all PAHs are carcinogenic. Consequently, a suitable index for carcinogenic potential of the types of PAHs in the air is needed. Currently, the presence of BaP may be provisionally regarded as a sufficiently active indicator of the carcinogenic potential of PAH.

A number of additional studies are needed to reveal the total health effects of PAHs. For example, PAH profiles detected in different emissions and workplaces sometimes vary greatly among one another and from PAH profiles in the ambient air. Considerably more data are necessary in order to develop a precise index for the carcinogenic potential of all PAH profiles, which can occur under conditions relevant for lung cancer risk estimates. In addition, the carcinogenic effects of PAHs may be affected by the synergistic effects of other VOCs emitted into the atmosphere.

Polychlorinated Biphenyls

Polychlorinated biphenyls (PCBs) are not found in nature but are prepared industrially by the catalytic chlorination of biphenyl and anhydrous chlorine. The resulting product is a complex mixture of chlorophenyls. Depending upon the chlorine atoms per molecule, the chlorophenyls may be solid, an oily fluid, or sticky resins. PCBs are characterized by their thermal stability, resistance to oxidation (nonflammable) acids, bases, and other chemical agents and their electric insulating properties.

Indoor PCBs come from a number of sources. A past common source was defective capacitors of luminous discharge lamps. PCBs were also commonly used as plasticizers, as sealants in building construction, surface treatment of textiles, waterproofing of wall coatings in paints and printers inks, and sealant putties.

Since 1973 the manufacture, sale, and use of PCBs have been prohibited in the 24 member countries of the Organization for Economic Cooperation and Development (OECD). However, due to widespread use of PCBs in buildings built prior to 1973, and because PCBs have a long life, they still represent an indoor pollution source. In recent years, high concentrations of PCBs in ventilation ducts indicate that air-conditioning systems remain a possible source of PCBs in air (Petreas et al., 1990).

Only a few studies have been made of the concentrations of PCBs in indoor atmospheres. In a study of a large school building constructed prior to 1971, indoor concentrations of more than 1,000 mg/m^3 were found (Burkhardt et al., 1990). In a study of 100 German buildings, emission rates were found to depend upon the different types of PCB products as well as atmospheric conditions. The highest emissions occur during summers and range from 500 mg/m^3 to 1,835 mg/m^3 (Balfanz et al., 1991).

PCBs have a number of health effects. They are absorbed by humans through the gastrointestinal tract, lungs, and skin. They are stored in the adipose tissues. There is also some placental transfer, and they are found in human milk. Several cases of chloracne, hyperpigmentation, gastrointestinal disturbance, elevated serum enzyme, metabolism abnormalities, and numbness of extremities have been reported among people highly exposed to PCBs (World Health Organization, 1993). There is also some evidence that PCBs may cause cancer.

Pesticides

Although elemental sulfur has been used for centuries as an insecticide to control ticks, thrips, chiggers, and mites, the modern era of pesticides has developed since 1940. As late as 1939, only 30 chemical pesticides were registered in the United States. In order to control pests and produce high-quality vegetables and fruits, pesticide use has ballooned. Although it is now estimated that 94 percent of the pesticide usage is for agricultural purposes, pesticide chemicals are widely used to combat indoor pests. When properly applied, many pesticides can be used and still maintain a healthy environment both outdoors and indoors.

Indoor chemical pesticides are universally used to control termite infestations in buildings. They are also used by individuals to prevent bites from mosquitoes, chiggers, flies, ticks, fleas, and other insects. House dust can be a major source of pesticides. In personal use, some products such as deodorant spray, hair spray, and shaving foam contain pesticides. Pest strips are widely used in the home. They contain DDVP (dichlorvors or vapona) as their principal chemical. DDVP has been classified as a human carcinogen. Pesticide chemicals that contaminate the air are also used in spray paints.

Because of the potential health dangers posed by pesticides, the Environmental Protection Agency in 1985 developed a methodology for determining pesticide exposure in the general population (EPA, 1990). The methodology, known as the Non-Occupational Pesticide Exposure Study (NOPES), was designed to estimate the exposure of the most commonly used household pesticides via air, drinking water, food, and dermal contact. In two cities studied (Jacksonville, Florida, and Springfield/Chicopee, Massachusetts), the average pesticides in the home numbered 4.2 and 5.3, respectively. Of the 34 pesticides studied, indoor air concentrations were substantially higher than outdoor air concentrations. There were also seasonal variations in concentrations. Concentration levels are dependent on a number of factors, including temperature, patterns of pesticide use, use of heating and cooling systems, and occupants' activities. Pesticides enter the human body in a number of ways. Food appears to be the major source for most pesticides in the body. Air is also an important source, whereas drinking water is of very limited importance.

Because pesticides are chemicals intended to kill insects, rodents, and other pests, they are inherently toxic, and without proper usage or storage they are potential health hazards. It is generally known that less than 50 percent of people read pesticide labels as to the proper application. Because of the great variety of pesticides, risks associated with use depends upon such factors as the concentration in the body, biological reaction of organs, differences in mode of action, and whether or not they can be eliminated from the body.

The most direct effect of pesticides on humans is poisoning. For example, in 1987 the poison control centers of the United States reported 57,430 cases of pesticide exposure, of which 98 percent were accidental. When there is high-dose exposure to certain pesticides, acute intoxication can result. There are also many studies of long-term, low-dose exposure of animals. However, the relevance of animal reactions may not be applicable to humans. Epidemiological confirmation is available for only some of the effects on humans (Maroni and Fait, 1993). Most studies have concentrated on the carcinogenic effects of pesticides and have not given sufficient attention to other health effects (see Tables 10 and 11).

TABLE 9
Effects of Formaldehyde on Humans after Short-Term Exposure

Effects	Formaldehyde Concentrations (μg/m³) Estimated Median	Reported Range
Odor detection threshold	0.1	0.06–1.2
Eye irritation	0.5	0.01–1.9
Throat irritation threshold	0.6	0.1–3.1
Strong sensation in nose, eyes	3.1	2.5–3.7
Lachrymation (30-minute tolerance)	5.6	5.0–6.2
Strong lachrymation (1 hour)	17.8	12.0–25.0
Life threatening; edema, inflammation, pneumonia	37.5	37.0–60.0
Death	125.0	60.0–125.0

Concentrations in μg/m³ = 0.813 ppm

Source: World Health Organization. *Air Quality Guidelines for Europe.* WHO Regional Publications, European Series No. 23. Copenhagen, Denmark: World Health Organization, 1987.

Allergies have been related to a number of pesticides. The importance of pesticides as to effects on human health requires more study.

Sources of Volatile Organic Compound Emissions

Volatile organic compounds are found in a large number of building materials and house furnishings (Hansen, 1991). VOCs are recognized as major contributors to indoor air contaminants. They are present due to different types of manufacturing processes and, to a lesser degree, methods of installation. Because of the great variations in potential VOCs and their production, emission levels vary greatly. In addition, the rate of emission from several sources is complicated, for there is a tendency for some materials to act as sinks, that is, VOCs released from one source may be absorbed into another and later rereleased into the air. The materials that emit VOCs are classified generally under surface adherents, manufactured wood products, and indoor furnishings.

Surface Adherents

Adhesives An adhesive is a substance applied to a surface in order to bind it to another surface. Adhesives are applied in a liquid or

TABLE 10
Health Effects of Prolonged Exposure to Pesticide

Pesticides	Evidence Well Established	Evidence Requiring Further Study
Herbicides		
Phenoxyherbicides 2,4,5-T; 2,4-D; MCPA, and related compounds (TCP, TCDD)	Chloracne	Cancer soft tissue sarcoma Hematopoietic systems stomach colon prostate Teratogenesis
Triazines	Lung cancer	Ovarian cancer
Arsenicals	Liver disease	—
Chlorinated Hydrocarbons		
Chlordane/Heptachlor disorders	—	Possible myelolymphoproliferate Brain cancer
Hexachlorobenzene	—	Porphyria Possible liver cancer
Dichlorodiphenyl-trichloroethane (DDT)	—	Chloracne Possible chromosome aberrations, high cholesterol and triglyceride levels, tremors, muscular weakness, neurotoxic effects
Synthetic pyrethoids	—	Reversible paresthesia
Organophosphorus esters	—	Delayed neuropathy, possible chromosome aberrations, central nervous system aberrations, non-Hodgkins lymphoma
Halogenated Hydrocarbons		
Dichloropropene	—	Myelolymphoproliferate disorders
Pentachlorophenol (PCP)	—	Hepatic effects Aplastic anemia
Dibromochloropropane (DBCP)	Spermatogenesis suppression	
Methylbromide	—	Possible mild neurotoxic effects
Phenoxy		
2.4, 5-T; 2, 4-D, and related compounds TCP and TCDD	—	Chloracne, possible soft-tissue sarcoma, lymphatic and hematopoietic system; stomach; colon; prostate teratogenesis
Triazines	—	Lung cancer, possible ovarian cancer
Arsenicals	—	Liver disease
Copper sulfate	—	Liver disease
EDB	Sperm abnormalities	—
Carbamates and Carbaryl	Chromosome aberrations Sperm abnormalities	— —
Orgonochlorine Insecticides		
Chlordane/heptachlor	—	Myelolymphoproliferate disorders

(continues)

TABLE 10
(continued)

Pesticides	Evidence Well Established	Evidence Requiring Further Study
Hexachlorobenzene	Porphyria	Brain cancer
DDT	Chloracne	Liver cancer Chromosome aberrations, high cholesterol and triglyceride levels Tremors Muscle weakness
Synthetic pyrethroids	Reversible paresthesias (skin sensations)	Neurotoxic effects —
Organophosphorus esters	Delayed neuropathy (some compounds only, often acute exposure)	Chromosome aberration Central nervous system aberrations
Copper sulfate	—	Non-Hodgkin's lymphoma Liver disease

Source: U.S. Environmental Protection Agency. *Introduction to Air Quality.* Report no. EPA/4003-91/003. Washington, DC: Environmental Protection Agency, 1991.

viscous state to a surface but on drying become solid. Most adhesives contain VOCs, the greatest emissions occurring during the drying process. Adhesives are generally classified according to the type of resin used as the base material.

Resins Resins (a class of solid or semisolid substances) are either natural or synthetic. Natural resins characteristically have low emission potential. In contrast, synthetic resins normally consist of organic compounds that have high organic levels of VOCs.

Sealants A sealant is a substance that is applied at openings in a building to eliminate penetration of liquid, air, and gas. The term *sealant* usually refers to the sealing of outdoor openings, whereas *caulk* is used for indoor applications. Sealants are used on a wide variety of surfaces such as glass, concrete, masonry, wood, plaster, and metals. Most of the contaminants from sealants are released upon installation or at the curing stage. The base resin is important in identifying the emission potential of a sealant, because it determines the type of liquid as well as the binders or modifiers. The actual hazard of a sealant depends on such factors as the toxicity and volatility of its ingredients, the quantity of the sealant used, and the curing time. Sealants are a definitive indoor air quality concern because of their potential VOC emissions.

Paints Paints are a major component in construction, used to protect surfaces against corrosion, weathering, and damage. All paints, stains, and varnishes require resins and oils to form a film that adheres to a surface. All coatings require liquid and organic substances to provide fluidity for application and adhesion. The amount of solids in a coating is a fundamental guide to potential VOC emission levels. A low-solid, organic, solvent-based paint will have a higher percentage of solvents compared to a higher solid-based product and, therefore, a higher emission rate. Conversely, the higher the solid content, the lower the emission of VOCs. The type of carrier used, water amount, and type of organic solvents further define the potential for VOC emissions. Recently, low-emitting stains and varnishes with low VOC emissions are manufactured more often than are similar paints (Holmberg, 1987).

Some water-based paints have low emissions. Most of these contain latex and water-soluble binders. Less typical are those that require an organic solvent additive, which raises the VOC emission levels. Water-based paints, however, require preservatives and fungicides such as arsenic, formaldehyde, and ammonia compounds. In 1991 NIOSH recognized them as chemical hazards but did not forbid their use in paints.

Manufactured Wood Products

Plywood Plywood consists of several layers of wood that are permanently bonded together by adhesives. Plywood has become a major construction material. Plies of soft- or hardwoods are stacked with the grains alternating at right angles, giving the composite board equal strength in length and width; there is less dimensional change in moisture content and less susceptibility to splitting. For interior finishes, plywood provides an economical alternative to more expensive solid woods.

The effect of plywood on air quality depends upon the bonding substances. Hardwood plywood of interior grade is typically bonded with urea-formaldehyde. Manufacturers are attempting to decrease the use of urea-formaldehyde in the production of plywoods, and emissions may be reducing. There is still some concern, however, when large amounts of plywood are used in interior construction. Building occupants must still rely on ventilation to reduce emission levels.

Particleboard Particleboard is a composite product made from wood chips or residues bonded together by adhesives under heat and pressure. The adhesive typically used in production is urea-formaldehyde. Particleboard is used for constructing floors as well as walls and roof sheathing. The most common use is for doors, cabinets, and a wide variety of furnishings, such has tables and chairs. It can be formed into many shapes. Particleboard is used because it is inexpensive, flat, and strong and has little warp. The major indoor pollutant problem of using particleboard is the emission of formaldehyde into the atmosphere. After installation, trace amounts of formaldehyde can be emitted for several years.

Indoor Furnishings

Household Items Many items found in households contain air pollutants. Foam in chairs, paint on walls, varnish on furniture, and dye in fabrics emit VOCs in varying amounts.

Room Partitions In many modern office buildings, prefabricated movable partitions (forming cubicles) are commonly used to change room configurations. Normally made of particleboard or plywood, partitions have a high potential to emit VOCs into the rooms due to the high square footage of material compared to room volume. Currently, there are few national or state regulations regarding emissions from building furnishings and equipment. The variety of VOC emissions that may be released in a room provides a complex and interactive environment. As a result, ventilating these rooms becomes critical in maintaining a satisfactory air quality standard.

Carpets In recent years carpets have received much attention for their effects on air quality and the degree to which they present a health hazard. The components of carpet include fibers, backing, adhesives, and, normally, a carpet pad.

Originally, fibers in carpets were primarily wool. Today, many carpets are produced from synthetic fibers such as nylon, olefin, polyester, and polyethylene terephthalate (PET). These are derived from petrochemicals and are used because of their overall durability and resistance to abrasion. There is no evidence that synthetic fibers are a problem in air quality. Carpet fibers are dyed with a variety of colors. Likewise, there is no evidence that modern dyes present a problem in air quality.

VOC emissions from carpets derive from the adhesives used to glue them to the pads or floors. The four most commonly used types of carpet adhesives are organic solvent–based styrene butadiene rubber (SBR), latex emulsion, acrylic latex, and hot-melts (ethylene vinyl acetate).

The effect of carpet adhesives on indoor air quality varies greatly depending upon the chemicals used, with VOC emissions ranging from 100 to 1,000 times greater than emissions from the carpet itself. Manufacturers are developing adhesives that reduce VOC emissions. The adhesive and sealant industries have formed a council in cooperation with EPA and the Carpet and Rug Institute (CRI) to study methods to reduce VOCs. The CRI has recommended that carpet installers using high-emitting adhesives install carpets only where there is excellent ventilation.

Carpets contain a backing that keeps fibers in place and strengthens dimensional stability and resilience. The primary backing product consists of polypropylene, which is resistant to acids and abrasions. A secondary backing, used to add strength and stability, is made from fabric, jute, or polypropylene and is bonded by a latex or polymer coating. The chemicals styrene and butadiene are known irritants to mucous membranes and skin. They also emit steady amounts of the by-product 4-PC, which has an offensive odor.

Wall Coverings The traditional wall covering is paper, but in recent years vinyls and fabrics have also been used. The installation of each of these materials requires an adhesive. For wallpaper, the adhesive is usually 100 percent water-based paste or natural starch, which do not possess air contaminants. When fabrics are used to cover walls there is a danger of contamination from formaldehyde, which is sometimes used in fabrics to improve water resistance and the color fastness of dyes.

Wall coverings made of vinyl have the highest potential to affect air quality. Vinyl covering has been known to emit such VOCs as amines, n-decane, formaldehyde, 1,2,4 trimethylbenzene, and xylenes (Brooks and Davis, 1992). The new vinyl covering may emit odor due to the printing and finishing solvents. A major concern in using vinyl covering is the sealing quality on the walls. In humid climates where there is no season when moisture evaporates, molds and mildews can grow behind the vinyl covering. Their fungal growth can result in odors and stains and ultimately create a health problem.

Acoustical Ceilings Acoustical tile has become a common ceiling covering in recent decades. Most acoustical tile today is produced from mineral or wood fibers compressed to the desired shape and thickness. The material used for acoustical tile varies greatly based on acoustic requirements, moisture resistance, fire protection, and aesthetic preference. For example, a textured surface provides better acoustical characteristics than does a smooth surface.

Air contaminants depend to a considerable degree on tile material. Some tiles are made from formaldehyde, and proper ventilation must be maintained to reduce the effects of VOC emissions. If a porous ceiling is damaged, VOCs may collect in the ceiling spaces and gradually emit into the room. When mineral fibers are used, the moisture contact may be high, encouraging the growth of biological fungi. Finally, fine particles may be emitted into the atmosphere over time from ceilings made of minerals, wood, or fiberglass.

Resilient Flooring Resilient flooring consists of tiles or sheets attached by adhesives to the floor. They are made primarily of polyvinyl chloride resins (PVCs), plasticizers to provide flexibility, fillers, and pigments for color. Rubber tile is made from a combination of synthetic rubber (styrene butadiene), pigments, extenders, oil plasticizers, and mineral fillers. Linoleum, first produced in the mid-1800s, is a natural organic and biodegradable product. The major components are linseed oil, resin, cork powder, pigments, and natural mildew inhibitors. Due to oxidation of the linseed oil, linoleum is naturally antibacterial and termite-resistant.

Typically, no single component of resilient flooring has high VOCs. However, the compositions of different types of sheets vary greatly, and so the VOC emission rates vary. Due to the typically large ratio of surface areas of vinyl and linoleum flooring to spatial volume, this product should not be installed without adequate ventilation (Saarela and Sandell, 1992). Low emission levels can continue for a long period, but VOC concentration levels normally decrease 24 hours after installation. Since the plasticizer contains the highest VOC emissions, more rigid, less-plastic vinyls are recommended. However, if low-emission flooring is glued to the floor with high-emission adhesives, the VOC rate will rise.

Noise Pollution

Noise is normally defined as unwanted sound. Atmospheric sound, a form of energy, is produced by waves that represent compression and decompression of molecules of air. The movement of the air molecules produces variations in normal atmospheric pressure. The human eardrums respond to these differences and begin to vibrate. The transmission of these vibrations to the inner ear and their ultimate interpretation by the brain result in the sensory perception of sound. Sound can be transmitted only through a medium that contains molecules. It cannot move through a vacuum. Sound also moves at different speeds through different types of media. For example, the speed of sound in air is about 343 meters per second (m/s; 1,125 feet per second), although it varies with air temperature; through steel it is 5,029 m/s (16,500 feet per second).

The sound characteristics most important to human hearing are frequency and intensity. Frequency is related to the rate of variation in pressure associated with the propagation of sound waves. If sound produces 1,000 complete wave oscillations per second, its frequency is 1,000 cycles per second or 1,000 Hertz (Hz), the commonly used reference unit. Sound is normally represented as a spectrum of frequencies. Within the range of frequencies, sound is normally characterized by the dominant frequencies. For example, humans can hear sounds in the frequency range of 50–20,000 Hz. Male speech is characterized by low frequencies, less than 2,000 Hz, whereas female speech is characterized by higher frequencies. All frequencies are not heard equally well. Humans hear best those frequencies that correspond to human speech (500–4,000 Hz).

The intensity of sound is a response to the energy level, or amplitude, of the sound waves. The amplitude (intensity) increases with the sound energy. Humans perceive increasing intensity as an increase in loudness (see Table 12).

Sound intensity is expressed in decibels (dB) and is determined by measuring the pressure of sound with an instrument called a sound pressure level meter. In normal usage, dB refers to sound pressure levels on a logarithmic scale relative to a reference pressure, which is defined as the threshold of human hearing, or 0.002 microbars. A microbar is one-millionth of normal atmospheric pressure.

Personal Sound Pollution

Sound pollution is an increasing problem in the home, at the workplace, as well as during other activities. In homes, sound pollution occurs largely as a response to the decibel level that music is played and, in recent years, relates most directly to rock. In the workplace, high decibel levels of sound are usually a response to operating machinery. Outdoor sound from jet airplanes and heavy trucks may penetrate the indoor environment. The increasing use of personal headsets, reducing disturbances to others, perhaps has encouraged an increase in the level of routine

TABLE 11
Health Effects on Humans of Short-Term Exposure to Selected Pesticides

Pesticides	Health Effects
Organophosphates	
Acephate, chlorpyrifos diazinon, dichlorvos (DDVP), disulfotion, malathion	Cholinesterase inhibitor
Pyrethroids	
Permethrin, tetramithrin	Possible human carcinogen
Carbonates	
Bendiocarb	Cholinesterase inhibitor
Propoxur	Possible human carcinogen Cholinesterase inhibitor
Carbaryl	Chromosome aberrations Sperm abnormalities Cholinesterase inhibitor
Chlorinated Hydrocarbons	
Chlordane, dicofol lindane (HCH; BHC)	Possible/probable human carcinogen contact exposure
Synergists	
MGK264, piperonyl butoxide	Inhalation toxic

Source: U.S. Environmental Protection Agency. *Introduction to Air Quality.* Report no. EPA/400/3-91/003. Washington, DC: Environmental Protection Agency, 1991.

decibel exposure to the listener. Anyone living or walking along a city street, upon hearing loud music blasting from a passing vehicle, can only wonder at the decibel level inside.

The decibel level affects humans in a number of ways. As the sound level rises, the first effect is likely to be psychological annoyance. The degree of annoyance may depend upon the activity in progress, previous conditioning, and the nature of the sound. Annoyance may also be related to whether the sound is continuous or intermittent. It is generally recognized that the higher frequency sounds, such as 5,000 Hz, are more annoying than lower frequencies at the same sound level.

Loud sounds can also interfere with human voice communication. Noise not only interferes with speech but also makes it difficult to understand verbal communication. As sound intensity increases, speech is less well understood. Consonants, most important in understanding words, are masked more than vowels. Depending upon the sound level and the source and relative frequency spectrum of speech, noises may mask some speech sounds but not others. A classification has been developed that rates noise level in its ability to interfere with speech communication. The Preferred Octave Speech-Interference Level (PSIL) is based on sound pressure levels centered on octave bands between 500 Hz and 2,000 Hz, which cover the human speech spectrum. This classification makes it possible to determine when speech communication is normal, difficult, or impossible.

Noise exposure is frequently considered to be responsible for a variety of physical as well as personal psychological problems. Of these health effects, impairment and loss of hearing are well proven. Noise-induced hearing loss is usually associated with industrial noise exposure and, more recently, with high-level sound on radio. The relationship between hearing loss and sound level is influenced by a variety of factors. These include sound intensity, exposure duration, sound frequency, individual susceptibility, susceptibility to specific frequencies, and characteristics of the time-dependent development in permanent threshold changes. There is growing evidence that noise adversely affects general health, particularly the cardiovascular system.

The U.S. Department of Labor's Occupational and Safety Health Administration (OSHA) and the Environmental Protection Agency have recommended permissible noise standards to protect individual workers from overexposure to noise levels that have the potential for causing hearing loss (see Table 13). The Department of Labor noise standards indicate that a worker could be exposed to an average of 90 dB for an eight-hour day for five days a week throughout a work career without risk of suffering a significant hearing loss. With increased sound levels, the permissible exposure period per day decreases.

The 1983 OSHA Hearing Conservation Amendment requires employers to administer hearing conservation programs whenever the eight-hour average noise in the workplace exceeds 85 dB.

Community Sound Pollution

Unlike indoor exposure, whether at home or work, community noise does not result in impairment of hearing in most instances, but it does affect the quality of human life by interfering with speech communication, disturbing sleep and relaxation, and intruding on privacy. Because of these annoyances, pollution control agencies receive numerous complaints from individuals. In

TABLE 12
Decibel Levels: Sources and Human Response (in decibels)

Source	dB	Response
	140	Painful to ears
	130	
Jet aircraft, rock music, and motorcycle (within 20 feet)	120	
	110	
Pneumatic drill, heavy truck (over 50 feet)	100	Creates permanent hearing impairment
	90	
Motor vehicle under 50 feet	80	
	70	Annoying
	60	Affects speech communication
Home	50	Quiet
	40	
Whisper	30	
Radio-TV studio	20	Very quiet
	10	Barely audible
	0	Threshold of hearing

Source: Compiled National Bureau of Standards, *Fundamentals of Noise Measurement Rating Schemes and Standards,* National Technical Information Service, U.S. Department of Commerce, Publication No. 300.15, Washington, DC, 1973.

TABLE 13
Occupational Noise Exposure Standards

Exposure Duration (hr/day)	Permissible Sound Level (dB)
8.0	90
6.0	92
4.0	95
3.0	97
2.0	100
1.5	102
1.0	105
0.5	110
0.25 or less	115

Source: Compiled from W. Burns and D. W. Robinson, *Hearing and Noise in Industry,* Her Majesty's Stationery Office, London, 1970.

residential areas, noise pollution is frequently associated with domestic activities such as noise from air conditioners, lawn mowers, blaring radios and TVs, barking dogs, playing children, and loud parties. These noises may be not particularly loud but can be most annoying.

Sound Control

Several federal agencies have authority to regulate and control noise: OSHA, EPA, the Federal Aviation Administration (FAA), the Federal Highway Administration (FHWA), the Federal Railway Administration (FRA), and the Department of Housing and Urban Development (HUD). The FAA has authority to impose and enforce noise emission standards for commercial aircraft operation; the FRA sets standards for railways. The FHWA is responsible for issuing standards for highway construction. OSHA has the responsibility to protect industrial workers from noise that impairs hearing. HUD's authority involves the establishment of noise control standards for HUD-assisted new housing construction.

The EPA has responsibility for administering the Federal Noise Control Act of 1972 and the Quiet Communities Act of 1978, both of which provide standards for control of community noise. The Noise Control Act of 1972 provides for the development of standards to identify levels of environmental noise based on the need for the protection of public health and welfare. This legislation also requires EPA to identify major noise sources that require regulation. Noise standards have now been established for interstate motor carriers and buses, medium- and heavy-duty trucks, railroad transportation, and construction equipment. The Noise Control Act of 1972 also authorizes regulation of manufacturers of a variety of equipment and appliances, requiring labeling of the sound levels their products produce.

The EPA, in administering the Quiet Communities Act of 1978, has provided assistance and financial aid to state and local governments and regional planning agencies to investigate noise problems, plan and develop a noise control capability, and purchase equipment. It has also given assistance to facilitate communities to develop and enforce community noise control measures, as well as to develop abatement plans for areas where noise is a potential problem, such as airports, highways, and railyards.

Building Pollutant Disease

Sick building syndrome (SBS) is the term applied when at least 20 percent of the people in a building experience symptoms of illness while the causal agent remains unknown. When one or more persons experience illness and the causal agent is known, the term is *building-related illness* (BRI).

Sick Building Syndrome

Sick building syndrome began to be widely recognized after the energy crisis of the early 1970s. In support of the effort to conserve energy, buildings throughout the world were better insulated. Although this conserved energy, the buildings became tighter, and the indoor air quality declined as the exchange of indoor air with outdoor air lessened.

Because of the magnitude of the problem, the major question focused on which contaminants existed in a building to cause illness. The answer has proved difficult and has eluded the medical profession. It has become evident that contaminants in a building are complex. Even when one or two contaminants are identified, another, unidentified contaminant can cause illness. It is entirely possible that some contaminants act synergistically to cause illness. There is a wide variety of symptoms that can occur singly or in combination. The illness may be nonspecific and often resembles a common cold or other respiratory disease.

The adverse health effects of SBS appear in a number of ways. A common complaint is eye irritation, a burning, dry, gritty sensation without evidence of inflammation. The severity of the irritation can vary from day to day and can disappear after the person is out of the building for several days.

Another common symptom is nasal disturbance. When a person enters a building a feeling of stuffiness develops but normally disappears when the person leaves. For many people, the stuffiness increases as the seasonal temperature rises. Other nasal symptoms are more variable and are apt to be less persistent. These reactions sometimes appear as nasal irritation.

The building contaminant may also affect the throat with persistent dryness, but rarely with any irritation or inflammation. This condition is usually relieved temporarily by drinking cold liquids. When the lower respiratory tract is attacked, the individual may experience a shortness of breath or an inability to breathe deeply, but it is not related to a lung infection or bronchial asthma. After a person is exposed for several hours to the indoor contaminants, a headache usually centers in the frontal lobe. It can be quite severe. Fatigue, dizziness, difficulty to concentrate, and general malaise are frequently cited by affected individuals. Another problem is dry skin. When the air is warm and dry, any air movement may cause skin to lose moisture. This will cause it to become rough and cracked. If this continues, skin rashes may result due to the greater exposure to contaminants.

There is a wide variety of factors, alone or in combination, that may cause or trigger SBS. More and more investigations indicate that microorganisms play an important role. A three-year study revealed a high correlation for visible mold growth and respiratory health problems. In a Finnish laboratory study, it was demonstrated that volatiles from microbial cultures have an impact on cilia cells in respiratory airways (Joki et al., 1993). In another Finnish study, a moldy odor in a daycare center revealed high possibilities for SBS symptoms (Jaakola et al., 1993).

In the United States, most of the information on building-related illnesses has derived from studies conducted by federal and state agencies on health hazards (Strom et al., 1993) rather than epidemiological studies. Most of the information has been gathered by questionnaires completed by individuals. As a result, there is considerable variation in the range of symptoms reported. Consequently, comparisons of symptoms from one place to another are difficult.

Building-Related Illnesses

The causes of BRIs can be determined. They are generally related to allergic reactions and infections. The allergies include Legionnaires' disease, humidifier fever, hypersensitivity pneumonitis, and rhinitis. Bacteria, fungi, and viruses are also causes of BRI infections. An evaluation of the symptoms provides an understanding for diagnostic activity.

Pontiac Fever A number of fevers are related to bacteria. In 1968, there was an outbreak of nonpneumonia fever in Pontiac, Michigan. The incubation period was from several hours up to 48 hours, and the illness disappeared in two to five days. This disease usually appears as an epidemic, and 95 percent of those who are exposed to this type of *Legionella* become ill (Dennis, 1990). However, no fatal cases of Pontiac fever have been reported.

***Legionella* Pneumophila** In addition to the *Legionella* that causes Pontiac fever, there are 29 other species. Of these, *Legionella pneumophila* is most important. It was first identified after an outbreak of pneumonia at the 1976 American Legion Convention in Philadelphia (Fraser et al., 1977). The incubation period is as little as two days up to 13 days after exposure. The disease creates an abrupt reaction: high fever, headache, muscle pains, and shivering followed by coughing and respiratory failure. Legionnaires' disease may also affect other organs and require special antibiotic treatment. About half of the affected persons become disoriented; 30 percent develop vomiting and diarrhea; more than 10 percent of persons affected will die as a result (Fraser et al., 1977).

The major source of *Legionella pneumophila* was not discovered until 1978 (Dondero et al., 1980), when it was found that the source of the bacteria was a ventilation system in a hospital. In that hospital, the main cooling tower failed and an auxiliary cooling tower, unused for two years, automatically came into operation. After three days, the first case of Legionnaires' disease was identified. It became evident the air from the newly vented system was the source of this bout of *Legionella pneumophila.*

Since then it has been recognized that a cooling tower provides an excellent environment for the growth of *Legionella* bacteria. The microorganisms become airborne through the movement of air during the process. The occupants of the building are thus exposed to the microorganisms through the air-conditioning system. Since the number of microorganisms reaches a maximum in the larger ventilation systems, the disease is usually associated with large buildings such as hotels, office buildings, and hospitals.

Although the microorganism, *Legionella pneumophila,* is fairly common in the environment, outbreaks of Legionnaires' disease are somewhat rare. It appears that the infective dose must be high; the means by which the organism is disseminated is also critical. Microorganisms in major ventilation systems can be controlled by chemically treating the water that circulates through the cooling tower.

Rhinitis Symptoms of rhinitis (inflammation of the mucous membranes in the nose) are those generally associated with asthma and increase with duration of exposure, but the symptoms lessen and disappear after the affected person leaves the building. Allergic responses occur in the upper and lower tracts of the respiratory system. Rhinitis allergens are usually associated with older and poorly maintained buildings. A common source is humidifiers, particularly cold-water spray humidifiers contaminated by microorganisms.

Humidifier Fever Humidifier fever symptoms are flu-like and include lethargy, neuralgic pain in the joints, muscular pains, and fever. Some individuals experience headaches and sudden weight loss. The disease has a unique time pattern. The systemic and respiratory symptoms appear with initial exposure and improve with continued exposure until they disappear. On reexposure the symptoms reoccur. For example, in a workplace setting, symptoms might appear on Monday, improve during the week, and disappear by Friday, but they will reoccur the following Monday upon reexposure. This pattern clearly distinguishes humidifier fever from hypersensitivity pneumonitis (see discussion below).

The cause or causes of humidifier fever have not been isolated. Some organisms have been identified during outbreaks but not positively determined to be the cause. Immunological investigations reveal the presence of antigens in humidifiers. In bronchial outbreaks, moisture from humidifiers usually reproduces the symptoms and physiological changes recognized in humidifier fever cases.

Although humidifier fever and hypersensitivity pneumonitis cause many similar symptoms, there are fundamental differences. In humidifier fever, the lung function is not impaired, whereas hypersensitive pneumonitis impairs lung functions. Humidifier fever appears to be triggered by comparatively low antigen levels. In contrast, hypersensitivity pneumonitis is associated with massive antigen exposure.

Hypersensitivity Pneumonitis This illness is essentially an allergen lung disorder. It occurs in different intensities. At initial exposure, symptoms increase. However, symptoms usually lessen within two to seven days after exposure. Malaise and muscle pain are almost always present. Headaches are also common. In the most severe cases there is shortness of breath, fever, chills, and dry coughing. Only a few individuals within a group will be affected by the disease. Individual susceptibility is suspected to be a key factor. Recovery is rapid when the causal agent is removed, and the disorder rarely reoccurs.

Control of Indoor Pollutants

The huge variety of indoor pollutants includes pollutants that are physical, chemical, and biological in origin. Indoor pollutants have increased greatly not only in concentration but also in quantities due particularly to new chemical products. Many have little effect on health, whereas others are major health hazards (and include carcinogens).

Measurement of Indoor Contaminants

There are many methods by which to measure contaminants of the indoor atmosphere. In order to ensure accurate measurements, the selection of equipment and methodology is critical. Other considerations include the choice of a meaningful time scale, the type of calibration of samples, and the analytical techniques chosen. There must also be an evaluation of the types of analytical techniques used. The Environmental Protection Agency and the National Institute of Occupational Safety and Health have established standards for the analysis of indoor air (EPA, 1978; American Public Health Association, 1977).

Most analytical methods measuring atmospheric contaminants require a sampling procedure combined with an analytical technique. Two basic types of measurement techniques are recognized. The first method requires continuous air movement through an instrument giving a direct reading. This method combines sampling an analysis in a single step. The second method is known as an integrated or grab-sample procedure. This measurement technique involves sampling air to measure the level of contaminants, which is followed by a chemical analysis of the sample in the laboratory.

There are now seven analytical methods to measure the quantity of pollutants in the atmosphere. They are electrical, electromagnetic, chemical, magnetic, thermal, gas chromatography, and radioactive. They can utilize direct reading field instruments or laboratory-based analysis.

Particle Detection

Particles are most commonly collected onto a filter and then analyzed gravimetrically. Particles as fine as 0.5 mm can be detected. There are also instruments available that provide a direct reading of particles without an intermediate step.

Optical Detectors

These instruments are widely used in industrial hygiene and in emergency response situations where color is important in the analysis. This technique is used indoors only when the contaminant is at a higher level. It relies on a chemical reaction between a contaminant and a reagent to produce a color that may be interpreted visually or optically to determine the concentration of the pollutant.

Atmospheric Instruments

Temperature Indoor air temperature is measured by a thermocouple, a bimetallic thermometer, or a liquid thermometer. In the United States, the Fahrenheit (F) system is used (32 degrees marks freezing and 212 degrees marks the boiling point at zero elevation). In most of the rest of the world, the Celsius (C) system is used (freezing is 0 degrees and boiling is 100 degrees at zero elevation).

Humidity Water vapor in the atmosphere is normally measured as relative humidity, that is, the ratio of water vapor in the air at a specific temperature in relation to the maximum amount of water vapor that the air could hold at that temperature. Relative humidity can be measured using a hygrometer or psychrometer.

Air Velocity Air movement is normally measured using analog or digital meters. Meters must be nondirectional in order to ensure that the air speed is measured. The unit of measurement for air velocity is feet per minute (fpm) or meters per second (mps). Some meters also contain scales for measuring static pressure in ducts. A mechanical air velocity meter kit can measure air velocities and differential pressures.

In measuring the direction of air movement, smoke trails or smoke candles release visible smoke that blends with air to aid in observing direction. These devices must be used with care, for they can cause fire and create pollution in a room that can be an irritant to the nose and throat.

Noise There are several available noise meters. For most measurements, a sound level meter will be adequate. In some instances where there are numerous sounds, an octave band analyzer is desired.

Ventilation

There are two types of ventilation of a building. Natural ventilation occurs when air moves freely between the inside and outside of a building. In contrast, a mechanical system forces an exchange of air from outside to inside, including a system to distribute the air.

Natural Ventilation The amount of air flowing into and out of a building through doors, windows, and cracks will depend upon air pressure and thermal conditions. The pressure difference is the basic control of air movement. But the flow through the openings will also depend upon the orientation of the openings to the wind direction, the size of the openings, local obstructions such as a wall, and wind-induced negative or positive pressures associated with the roof or building edges. Open windows or doors on opposite sides of a building provide excellent conditions for natural ventilation. The natural exchange ventilation rate of a house will vary greatly, from less than 0.1 air changes per hour to several ACHs.

There are advantages and disadvantages to natural ventilation in a building. A major advantage is that there are no investments required in ventilation equipment. It can be controlled to a limited extent by opening and closing doors and windows. The major disadvantages are that it varies with weather conditions; the recovery of energy costs from the movement of air outdoors is also not possible. Under breezy and windy conditions a building may be subject to drafts. During the summer season, the temperature of a building may increase as outdoor air enters it; in contrast, during the winter season a building may be cooled from the outdoor air. Thermal comfort is therefore difficult to control.

There have been no empirical studies quantifying the effectiveness of natural ventilation in reducing or increasing indoor contaminant levels. In most instances, effectiveness is inferred from occupants in smell or comfort standards. For example, in homes where formaldehyde was a recognized contaminant, more than 80 percent of the occupants indicated that health symptoms diminished when windows were opened (Godish, 1989). There is, however, no evidence that natural ventilation is a way to reduce air contaminants. The air movement in natural ventilation cannot be controlled, thus its efficiency as an exchanger of air varies greatly.

Mechanical Ventilation Responding to so-called sick building complaints, the National Institute of Occupational Safety and Health concluded that the cause of illness in at least 50 percent of the cases was due to inadequate ventilation. Since most of the buildings in the past quarter-century have been designed to conserve energy, natural ventilation is at an absolute minimum. As a consequence, it is now recognized that a system of mechanical ventilation is necessary to change air. The primary purpose of mechanical ventilation is to provide a healthy and comfortable indoor environment for occupants.

In a natural ventilation system, when outdoor air is exchanged for indoor air, the indoor heat is lost. In order to overcome this disadvantage, a technique has been devised to recover energy in an air-to-air heat exchange. In this process, roughly equal amounts of air are moved from outside to inside. In an air-to-air heat exchange, heat is transferred from the warmer to the cooler airstream, ideally without any mixing between the two airstreams. Due to the transfer of heat, outdoor air is preheated in the winter and precooled in the summer. There are a number of types of air-to-air heat exchangers. In some, only heat is transferred between airstreams (these are usually referred to as "sensible exchangers"). In other exchangers, both heat and humidity are transferred between airstreams (Fisk and Turiel, 1983).

In large buildings, such as offices, hospitals, and auditoriums, mechanical ventilation systems are designed to provide for heating, ventilation, and air-conditioning (HVAC). For purposes of heating and cooling, a building may be divided into zones to serve different occupancy needs. For example, during occupied hours, the core zone, due to heat produced by light, people, and other sources, may require cooling; during unoccupied hours, and just before occupancy, heating may be necessary. The system must also be able to adjust to changing weather conditions.

In the design of a mechanical ventilation system, the rate of air exchange of pure air for pollutants is of major importance. This presents a difficult technical problem, for there is potentially a large mixture of pollutants. In addition, the system must be cost effective, which usually means that energy use is efficient and flexible enough to accommodate future changes.

Because movement of air comes from all parts of a room or building, there are problems associated with mechanical ventilation. One is that many of the mechanical ventilation systems have not been properly designed. For example, in response to escalating energy costs, building managers responsible for the operation and maintenance have modified the operation of HVAC systems, particularly in regard to percentage of recirculated and outdoor air used. Many HVAC systems operate totally on recirculated indoor air. Under these conditions, it is almost axiomatic that a building will have ventilation problems that increase air pollutants.

Another problem is that the air circulation does not reach all portions of the occupied space. In addition to the basic problem of ventilation efficiency, inadequate ventilation may occur as a result of imbalance in a system. When the HVAC system is imbalanced, some spaces receive more air than needed while others receive less. System imbalance may be due to poorly operating valves or dampers, dirty filters, improperly sized fans, inadequate provision for return of outside air, and significant differences between building spaces.

Control of Particulate Contaminants

Particulate matter is the most common air contaminant. The particles of greatest interest include such respirable particulates (10 microns or less in size) as:

1. Dust
2. Asbestos fibers
3. Allergens (pollen, fungi, and others)
4. Pathogens (bacteria and viruses) usually found on other particulate matter
5. Tobacco smoke

Indoor particulates are generated from many sources, including indoor as well as outdoor sources. The type and design of a filter determines its efficiency to remove particles of a given size and the amount of energy required in the process. ASHRAE has devised a standard rating known as Dust Spot and Anestance, which measures different aspects of performance.

Low-efficiency filters (ASHRAE Dust Spot efficiency rating of 10–20 percent or less) are used to remove the largest particles such as lint. The medium-efficiency filters (ASHRAE Dust Spot efficiency rating 30–60 percent) are much more efficient, removing bacteria, pollen, insect feces, dust, and so on. The high-efficiency filters (ASHRAE Dust Spot efficiency rating of 85–95 percent) are used to remove the finest particles.

The two basic types of filters to remove particulates from the atmosphere are mechanical and electrostatic.

Mechanical Filters Mechanical filters remove particles from air due to physical forces imposed on the particles by an airstream and filter. This technique is based on the concept that when a particle collides with the filter, it loses velocity and is captured. The filter performance is determined by a number of factors such as filter thickness, diameter, density, and air flow rate. Of these, filter thickness is of special importance. Filtration efficiency for the same size particles increases as filter thickness increases. However, an increase in filter thickness results in an increase in resistance to air flow. The pressure drop is directly proportional to filter thickness and inversely related to filter diameter. The effects of a decrease in air pressure is to reduce the actual rate of air flow through the filter and thus the volume of air cleaned. To compensate for a pressure drop, it is necessary to have stronger fans with a higher horsepower.

Electrostatic Filters Electrostatic filters remove solid and/or liquid particulates from gases by imposing negatively charged particles, which are then collected on positively charged plates. The electrostatic preregulation was first used in 1907 to collect mist from sulfuric acid plants. Since then, it has also been used to collect ferrous and nonferrous metal particles in various metallurgical processes. Today the largest single application of electrostatic precipitation is the collection of flash from coal-fired electric power boilers.

Electrostatic precipitation systems are widely used to collect dust because they have a high efficiency for all particle sizes. Ninety-nine percent efficiency can be achieved on even very large gas flows. They also have low operating and power requirements, because energy is only necessary to act on particles collected. Disadvantages include high capital costs and large space requirements. They are thus used exclusively at industry sites.

There are three types of fibrous media filters for indoor air cleaning. The dry filter has a high porosity and is normally used to collect such particles as dust. The viscous media filter consists of coarse fibers coated with an oily material to which particles adhere. Similar to the dry filter, the viscous fibers have a high porosity and a low resistance to air flow. They have high efficiency in collecting dust and lint. Renewable media filters automatically provide a fresh surface by use of a sensor when the previous surface becomes clogged. Renewable filters come in dry and viscous types and have medium-to-high efficiency of removal of contaminants.

Control of Gaseous Combustion Products

A wide variety of pollutants can be emitted from indoor combustion appliances and tobacco smoking, including carbon monoxide, carbon dioxide, nitrogen dioxide, sulfur dioxide, formaldehyde, and respirable particles. This section concentrates on the content of carbon, nitrogen and sulfur dioxides, and carbon monoxide.

Carbon Monoxide Carbon monoxide is another by-product of fossil-fuel combustion. Other major sources of indoor CO are tobacco smoke and vehicle engines running in garages. The CO seeps into the house and may collect to dangerous levels unless removed by a ventilation system.

A number of laboratory studies have been conducted in an attempt to reduce CO generation. This does not appear possible because of the complexity of the combustion process. There has been some success in reducing CO in the home, but in the process NO_2 emissions have been increased. Thus, all flame modifications must be carefully evaluated to determine their effect on emissions of all pollutants.

Carbon Dioxide Indoor carbon dioxide comes from three sources: CO_2 in outdoor air, the breathing process of humans and animals, and, most important, combustion processes. Of these, there is little possibility of reducing CO_2 in buildings. Since CO_2 is nonreactive, control of excess CO_2 concentration in the air is best done by ventilation.

Nitrogen Dioxide Nitrogen dioxide is produced when nitrogen and oxygen combine in any process where the temperature exceeds 1,800° Kelvin (Celcius + 273.15°). Such temperatures are reached in the flames of methane and kerosene combustion appliances. The emission from these sources can, however, be reduced. Many appliances have a pilot light to ignite the burner. Pollution concentrations can be reduced if the standing pilot is replaced by other ignition devices. A second procedure is to make burner modifications. The use of screens over flames to conduct heat away from the hottest portion of the flame reduces NO_2 production. In this procedure, however, carbon monoxide is produced. With proper screens it has been shown that CO production can also be greatly reduced. It has also been determined that a hood on a combustion stove can vent the NO_2 to the outside air.

Sulfur Dioxide Sulfur dioxide is also produced by combustion processes from fossil fuels with high sulfur contents. As a consequence, the concentration of SO_2 indoors can be significant. Indoor sulfur dioxide can be controlled by source elimination or the substitution of a sulfur-free fuel. Because the sulfur contents of many fuels are not revealed by the retailer, users may have to have the fuel tested by a laboratory.

Sulfur dioxide is partially removed indoors by absorption on such surfaces as wallpaper, paint, and carpeting materials. Increasing the natural removal of SO_2 indoors can reduce its concentration slightly. The best way to remove indoor sulfur dioxide is ventilation.

Control of Gas Phase Contaminants

Atmospheric contaminants can be removed from the air by adsorption or absorption processes.

Adsorption Many gases and liquid contaminants that come in contact with a solid surface will adhere to it. This process is known as adsorption. The solid surface is called the adsorbent; the material adsorbed on the solid is the adsorbate. Physical adsorption is due to forces of attraction between molecules of the adsorbate and molecules of the adsorbent. Thus, the adsorbed material rests on the surface and may be removed by scraping (Turk, 1977).

Since this type of adsorption is a physical process, no chemical reaction occurs. Nevertheless, heat is released. The amount of heat liberated is approximately equal to that released when the adsorbed gas or liquid undergoes condensation. The adsorption process is thus exothermic.

A number of types of materials can be used as adsorbents. These include activated carbons, molecular sieves, zeolites, porous clay minerals, silica gel, and activated alumina. Common to these materials are high surface area-to-volume ratios. They are characterized by vast submicroscopic pores and minute channels.

Of the adsorbents, activated carbon is the most common solvent used for air cleaning. Such raw materials as hardwood, coals, fruit pits, coconuts, and others are heated to a high temperature in an oxidizing process. The resulting material is highly porous with extended surfaces. Activated carbons thus have a high adsorbability. There is also high retention of gases and liquids.

The adsorption process is most practical for large buildings such as hospitals or industrial workplaces.

Absorption In the absorption process, the pollutant must be transferred from the atmosphere to a liquid. The washing liquid employed for such "air washing" depends on the type of pollutant to be removed. Water is commonly used, not only because of its low cost but because it absorbs many contaminants. To increase the efficiency of water, additives are occasionally used. The solubility of the fluid is critical. Gas scrubbing is not effective on contaminants, such as radon, that are not soluble. Because the washing liquid becomes saturated with pollutants, the washing liquid must be replaced to be effective. In addition to removing gaseous pollutants, some scrubbers are designed to remove physical particles.

The efficiency of air washing is determined by a number of factors, including the surface area for transfer, the flow rates and directions of air and washing fluid, and the driving force of the air that is forced to absorb the pollutants. The total rate of pollutant removal from a room will depend not only on the pollutant removal efficiency but also on the airflow rate through the air-cleaning device.

The capital and maintenance costs of air washing equipment are significant because of its complexity. The cost of energy is also high, but in winter the indoors can utilize the heat thrown off by such equipment. As a consequence, gas scrubbers traditionally have been used in large buildings, such as hospitals and offices, to remove air pollutants. In recent years, smaller models have been designed and are now available for smaller buildings.

Control of Volatile Organic Compounds

There are a number of processes and techniques that control volatile organic compounds.

Molar Ratio Molar ratio determines the bond strength and amount of swelling in particleboard or plywood. The reduction of this molar ratio (one gram per liter) of urea in formaldehyde can substantially reduce emissions. Nevertheless, although low molar ratios reduce the initial amount of formaldehyde emission, resins of low molar ratio are more vulnerable and create other problems in using synthetic wood products.

Process Modifications One of the best ways to control VOC emissions is to alter the manufacturing process. For example, many types of resins are injected into wood materials to produce plywood and particleboard, each of which has a different emission rate.

Resin Modification A variety of resins can be used to produce a formaldehyde product. Each of these resins have a different rate of emissions. Thus, the selection of a specific resin reduces the formaldehyde emissions from the UF foam.

Aging The emission of contaminants from volatile organic chemicals is normally rapid immediately after their use and decreases with age. For example, the storage of pressed wood products usually reduces the formaldehyde emission. However, some UF foams require a long period to deteriorate, thus the emission rate decreases with time. Knowledge of the UF foam resin used is necessary to determine when emissions are greatest.

Coating and Barriers Another means to reduce formaldehyde emissions from pressed wood is to coat surfaces. The coatings, which can be scavengers that react with formaldehyde, include polyurethane, lacquer, and paint. Standard latex paints are not effective. If the surface contains urea before painting, the emissions are reduced. Other coatings include melamine-impregnated paper, curing lacquer, decorative laminate, veneer, polyacrylamide, vinyl wallpaper, vinyl carpet, and floor coverings.
Ammonia Fumigation When formaldehyde materials are exposed to ammonia in the atmosphere, the emission rate is reduced. In laboratory studies, it has also been demonstrated that ammonia fumigation can reduce the formaldehyde emission rate to one-tenth or less of initial values. In all tests, the emission rates are reduced when the temperature and humidity are increased (Gesser, 1981).
Quality Control Tests have demonstrated that quality control of volatile organic chemicals in manufacturing processes is an important factor in reducing emissions. It has been proved that quality control of urea-formaldehyde is a major factor in formaldehyde emissions.

Control of Radon

Establishment of National Goals In 1985, after the discovery the year before of high radon levels in a house in Boyerstown near Reading, Pennsylvania, the Environmental Protection Agency established the Radon Action Program to provide technical assistance to states and to coordinate federal radon activities. The program includes:

1. An assessment of national and regional radon problems, including methods for predicting potential high-risk areas
2. Standardization of radon measurement methods
3. A system for evaluating the competence of firms and laboratories offering measurement services
4. Cost-effective techniques for reducing radon levels in existing and new houses
5. Programs for transferring technical assistance capabilities to states and the private sector
6. Action-level guidelines for reducing radon in houses
7. Effective communication levels to identify health concerns and recommend solutions to the public

In 1988, the U.S. Congress passed the Indoor Radon Abatement Act, which established national long-term goals with respect to radon levels in buildings, so that air within U.S. buildings would be as free from radon as the ambient air outside. This act requires EPA to:

1. Revise the "Citizens' Guide" [to radon detection and control]
2. Develop model codes
3. Provide technical assistance to the public
4. Establish a state grant program
5. Establish three regional training centers
6. Study radon in schools more extensively
7. Authorize a federal building study

Measurement Techniques Since radon is colorless and odorless, it can be detected only by physical measurement. EPA has evaluated a number of measuring techniques through the National Radon Measurement Proficiencies (RMP) program. This measurement technique includes:

1. Activated charcoal adsorption (ac) and charcoal loaded liquid scintillators (CLS)
2. Electric ion chambers (EIC)
3. Alpha-trick detectors (ATD)
4. Continuous radon monitors (CRM)
5. Grab-sample radon monitors (GRM)
6. Continuous working level monitors (CWLM)
7. Grab-sample working level monitors (GWLM)

AC, CLS, EIC ,and ATD are passive detection systems using no power sources. The radon enters and leaves the detector by diffusion. Such monitors require of additional reader systems to convert the information collected by the passive detectors into radon concentrations or working level units. The other detector types, CRM, GRM, CWLM, and GWLM, are called active detectors because they use the same type of pumping system to pump radon-containing air into or through the detector. The radon detector itself may be powered so that output readings can be obtained during data collection or immediately at the end of the monitoring period.

The AC, CLS, EIC, and ATD detectors measure radon-222 concentrations and then give results in terms of radon-222 concentration in picocuries per liter of air (pci/l) or becquerels per cubic meter of air (Bq/m^3). The CRM, GRM, CWLM, and GWLM detectors also monitor the concentration of radon-222 progeny in the air and give results in units or *working levels* (WLs), a term originally developed to compute radon progeny exposure to miners. It is generally assumed under normal conditions in a home that one working level of exposure is associated with a radon concentration of 200 pci/l. This number assumes that the concentration of radon-222 progeny in the air is 50 percent of its equilibrium value.

Radon Control Although no one knows how much household radon constitutes a hazard, the Environmental Protection Agency has set the safe level at 4 picocuries. This was an arbitrary decision, for it is relatively easy to reduce the level to 4 picocuries in all buildings, but it is extremely difficult to reduce the level to 1 picocurie. As more information is acquired, this standard will be adjusted either upward or downward.

After radon is detected in a building it is easy to control. When the building has high radon levels, two things need to be done to control the radon. The basement needs to be sealed so that no gas can permeate the walls from the outside; vents must also be installed in the basement to direct the radon outside. Ventilation is the key to indoor control. If ventilation is not developed, the concentration of radon will gradually increase. In drafty, old houses the danger from radon is greatly reduced. In contrast, modern houses are highly insulated in order to prevent the escape of heat and, of course, radon.

References

American Public Health Association. *Methods of Air Sampling and Analysis.* Washington, DC: APHA Intersociety Committee, American Public Health Association, 1977.

Aronow, F. S., J. Ferline, and F. Glauser. "Effect of Carbon Monoxide on Exercise Performance in Chronic Obstructive Pulmonary Disease." *American Journal of Medicine* 63 (1977): 904.

Ashford, N. A., and C. S. Miller. *Chemical Exposures: Low Levels and High Stakes.* New York: Van Nostrand and Reinhold, 1991.

Balfanz, E., J. Fuchs, and H. Kieper. "Erfahrungen mit Innenraumluffuntersuchungen auf Polycholorierte Biphenyle (PCB)," in Zusammenhang mit Dauereslastischen Dichtungsmassen. Meeting in Augsburg, November 1991, Hutzinger/Fiedler, eds., *Organohalogen Compounds* 7 (1991): 201–214.

Brooks, Braford O., and William F. Davis. *Understanding Indoor Air Quality.* Boca Raton, FL: CRC Press, 1992.

Brown, W. P., and M. Jones. "Mortality and Industrial Hygiene Study of Workers Exposed to Polychlorinated Biphenyl." *Archives of Environmental Health* 36 (1990): 120–129.

Burkhardt, U., M. Bork, E. Balfanz, and J. Leidel. "Innenraum belastung durch Polychlorierte Biphenyle (PCB)," in *Dauerelastischen Dichtungsmassen-Off. Gesundh-Wes* 52 (1990): 567–574.

Cameron, P., et al. "The Health of Smokers and Nonsmokers' Children." *Journal of Allergy* 42 (1969): 336.

Colley, J. R. T., et al. "Influence of Passive Smoking and Parental Phlegm on Pneumonia and Bronchitis in Early Children." *Lancet* 2 (1974): 1031.

Commins, B. T. *The Significance of Asbestos and Other Mineral Fibers in Environmental Ambient Air.* Maidenheath, England, 1985.

Dennis, P. J. L., "An Unnecessary Risk: Legionnaires' Disease," in P. R. Morey, J. C. Feeley, and J. A. Otten, eds., *Biological Contaminants in Indoor Environments.* Philadelphia: ASTM, 1990, pp. 84–95.

Dondero, T. J., et al. "An Outbreak of Legionnaires' Disease Associated with a Contaminated Airconditioning Cooling Tower." *New England Journal of Medicine* 302 (1980): 365–370.

Elkins, R. H., D. Y. C. Ng, J. Zimmes, and R. A. Macriss. "A Survey of Carbon Monoxide and Nitrogen Dioxide Levels in the Indoor Environment." Paper presented at 67th annual meeting of the Air Pollution Control Association, Denver, CO, 1974.

Fisk, W. J., and I. Turiel. "Residential Air-to-Air Heat Exchangers: Performance, Energy Savings and Economics." *Energy and Buildings* 5 (1983): 197–211.

Fraser, D. W., et al. "Legionnaires' Disease: Description of an Epidemic Pneumonia." *New England Journal of Medicine* 297, no. 22 (1977): 1189–1197.

Gesser, H. D. *A Study of the Reaction of Ammonia Gas with Urea-Formaldehyde Foam Insulation.* Winnipeg: University of Manitoba, National Research Council, 1981.

Godish, Thad. *Indoor Air Pollution Control.* Chelsea, MI: Lewis Publishers, 1989.

Hader, S., J. Kuhr, and R. Urbanek. "Sensibilisierung auf 10 wichtige Aeroallergene bei Schulkinderen Monatscher." *Kinderheilkd* 138 (1990): 66–71.

Hansen, Shirley. *Managing Indoor Air Quality.* Liburn, GA: The Fairmount Press, 1991.

Hirayama, T. "Non-Smoking Wives of Heavy Smokers Have a Higher Risk of Lung Cancer: A Study from Japan." *Journal of British Medicine* 282 (1983).

Holmberg, K. "Indoor Mold Exposure and Health Effects." *Proceedings of the 4th International Conference on Indoor Air Quality and Climate, Berlin* 1 (1987): 637–642.

Jaakola, J., et al. "Home Dampness and Molds as Determinants of Respiratory Systems and Asthma in Pre-School Children." *Journal of Experimental Analysis of Environmental Epidemiology.* Supplement 1, no. 3 (1993): 129–142.

Joki, S., et al. "Effect of Microbial Metabolites on Ciliary Functions in Respiratory Airways." *Indoor Air '93, Proceedings* [Helsinki, Finland] 1, no. 1 (1993): 259–263.

Kjaergard, S., L. Molhave, and O. F. Pedersen. "Human Reactions to Indoor Air Pollution: n-decane," in B. Seifert et al., eds., *Proceedings of the 4th International Conference on Indoor Air Quality and Climate, Berlin* 1 (1987).

Lioy, P. J., et al. "Persistence of Peak Flow Decrement in Children Following Ozone Exposure Exceeding the National Ambient Air Quality Standard." *Journal of the Air Pollution Control Association* 35 (1985): 1068–1071.

Lippmann, M. "Health Effects of Ozone: A Critical Review." *JAPCA* 39 (1989): 672.

Lowenstein, H., et al. "Indoor Allergens." *Journal of Allergy Clinical Immunology* 78 (1986): 1035–1039.

Maroni, M., and A. Fait. "Health Effects in Man from Long-Term Exposure to Pesticides: A Review of the 1975–1991 Literature." *Toxicology* 78 (1993).

Maroni, M., B. Seifert, and T. Lindvall, eds. *Indoor Air Quality.* Amsterdam: Elsevier, 1995, chap. 3.

Marrinowski, F., "Nationwide Survey of Residential Radon Levels in the U.S." *Proceedings of the Fifth International Symposium on the Natural Radiation Environment, Salzburg, Austria* 45 (1992): 419–424.

Martinowski, F. "Nationwide Survey of Residential Radon Levels in the U.S." *Proceedings of the Fifth International Symposium on the Natural Radiation Environment, Salzburg, Austria* 45 (1992): 419–424.

McGown, M. "The Trouble with Vinyl." *The Construction Specifier* (March 1992): p. 54.

Miller, E. Willard, and Ruby M. Miller. *Environmental Hazards: Radioactive Materials and Waste.* Santa Barbara, CA: ABC/CLIO, 1990, p. 5.

Morey, P. R., et al. "Environmental Studies in Moldy Office Buildings: Biological Agents, Sources, Preventative Measures." *Annals ACGIH* 10 (1984): 21–36.

Moschandreas, D. J., J. W. C. Stark, J. E. McFadden, and S. S. Morse. *Indoor Air Pollution in the Residential Environment.* Report No. EPA-600/7-78-229a, vol. 1: Data Collecting, Analysis and Interpretation. Cincinnati: EPA, 1978.

National Bureau of Standards. *Fundamentals of Noise Measurement Rating Schemes and Standards.* NTIS Publ. No. 300.14. Washington, DC: Government Printing Office, 1973.

National Research Council (NRC). *Asbestiform Fibers: Nonoccupational Health Risks.* Washington, DC: National Academy Press, 1984.

———. *Indoor Pollutants.* Washington, DC: National Academy Press, 1981.

Occupational Safety and Health Administration. "Occupational Exposure to Formaldehyde in Final Report." 29 CFR Part 1910.1048 (May 27, 1992).

Petreas, M. X., B. Renzi, D. Wijekoon, W. M. Draper, and R. D. Stephens. "PCB Contamination in the Office Building." 10th Dioxin Meeting, Bayreuth, Federal Republic of Germany, 1990.

Pollart, S., M. D. Chapman, and T. A. E. Platts-Mills. "House Dust Mite and Dust Control." *Clinical Review of Allergy* 6 (1988): 23–33.

Rabinowitz, M., G. Wetherill, and J. Kopple. "Absorption, Storage, and Secretion of Lead by Normal Humans." *Proceedings of University of Missouri Annual Conference on Trace Substance Environmental Health* 9 (1975).

Radian Corp. *Catalog of Materials and Potential Sources of Indoor Air Emission,* vol. 1. Research Triangle Park, NC: Radian Corp. 1 (1993): 4–19 (prepared for EPA).

Rosen, F. L., and A. Levy. "Bronchial Asthma Due to Allergy to Tobacco Smoke in an Infant." *Journal of the American Medical Association* 144 (1950): 620.

Saarela, K., and E. Sandell. "Comparative Emission Studies of Flooring Materials with References to Nordic Guidelines." *Proceedings of 1A2 '91* ASHRAE CBR, Health Building Conference, Washington, DC (Cital-ERF Topic I-9652, pp. 6–7), 1992.

Sawyer, R. N. "Indoor Air Pollution: Application of Hazard Criteria." *Annals of the New York Academy of Science* 330 (1979): 579–586.

Schaefer, K. E. "Physiological Stresses Related to Hypecopnia During Patrols on Submarines." *Undersea Biomedical Research Submarine Supplement* (1979): 515–547.

Schou, C., U. G. Svedsen, and H. Lowenstein. "Characterization of the Major Drug Allergen, Canafel." *Clinical and Experimental Allergy* 21 (1991): 321–328.

Spengler, J. D., B. S. Ferris Jr., D. W. Dockey, and F. F. Speizer. "Sulfur Dioxide and Nitrogen Dioxide Levels inside and outside Homes and the Implications on Health Effects Research." *Environmental Science and Technology* 13 (1979): 1276.

Strom, G., et al. *Microbial Volatile Organic Compounds (MVOC),* in American Society of Heating, Refrigerating, and Air-Conditioning Engineers, *International Conference on Building Design, Technology, and Occupant Well-Being,* 1993, pp. 351–357.

Tager, I. B., et al. "Effect of Parental Cigarette Smoking on the Pulmonary Function of Children." *American Journal of Epidemiology* 15 (1979): 110.

Turk, A. "Adsorption." In *Air Pollution,* vol. 4, *Engineering Control of Air Pollution,* 3d ed. Ed. A. C. Ster. New York: Academic Press, 1977, pp. 329–363.

U.S. Consumer Product Safety Commission. *Hazard Assessment for Pollutants Emitted During Use of Kerosene Heaters: Kerosene Heater Briefing Package.* Washington, DC: The Commission, 1983.

U.S. Environmental Protection Agency (EPA). *Air Quality Criteria for Lead.* EPA-600/8-77-017. Washington, DC: EPA, 1977.

———. *Air Quality Criteria for Oxides of Nitrogen.* Final Report, Report No. EPA 600/8-82-026F. Research Triangle Park, NC: EPA 1982.

———. *Air Quality Criteria for Ozone and Other Photochemical Oxidants.* EPA-600/8-78-004. Washington, DC: EPA, p. 19.

———. *Air Quality Criteria for Particulate Matter and Sulfur Oxides,* vols. 1–3. Report No. EPA-600/8-82-029, a–c. Research Triangle Park, NC: EPA, 1982.

———. *Assessment and Control of Indoor Air Pollution,* vol. 2. Report No. EPA 400/1-89-001C, Report to Congress on Indoor Air Quality. Washington, DC: U.S. Environmental Protection Agency, 1989.

———. *The Inside Story: A Guide to Indoor Air Quality.* Prepared by the U.S. Consumer Product Safety Commission. Washington, DC: EPA, 1995, 36 pp.

———. *Model Standards and Techniques for Control of Radon in Residential Buildings.* Report No. EPA 402 R 94 009. Washington, DC: March 1994.

———. *Nonoccupational Pesticide Exposure Study.* Report No. EPA/600/3-90/003. Research Triangle Park, NC: EPA, 1990.

———. *Quality Assurance Handbook for Air Pollution Measurement Systems—Principles,* vol. 1, and *Quality Assurance Handbook for Air Pollution Measurement Systems—Ambient Air Specific Methods,* vol. 2, Research Triangle Park, NC: Office of Research and Development, 1975 and 1977.

———. *The Senseless Killers.* Prepared by the U.S. Consumer Product Safety Commission. Washington, DC: 1993.

———. *What You Should Know about Combustion Appliances and Indoor Air Pollution.* Prepared by the U.S. Consumer Product Safety Commission and the American Lung Association. Washington, DC: 1993, 19 pp.

Wayne, W. S., et al. "Oxidant Air Pollution and Athletic Performance." *Journal of the American Medical Association* 199 (1967): 901–904.

Wilson, K., J. C. Chuang, and M. R. Kuhlman. "Sampling Polycyclic Aromatic Hydrocarbons and Related Semivolatile Organic Compounds in Indoor Air." *Indoor Air* 4 (1991): 513–521.

Wood, R., et al. "Antigenis Analysis of House Dust Samples." *American Review of Respiratory Diseases* 137 (1988): 358–363.

World Health Organization (WHO). *Man-made Mineral Fibers.* Environmental Health Criteria No. 77. Geneva, Switzerland: WHO, 1988.

———. *Polychlorinated Biphenyls and Triphenyls,* 2d ed. Environmental Health Criteria No. 104. Geneva, Switzerland: WHO, 1993.

Laws and Regulations

Although indoor pollution has long been recognized as a problem in industry, the problem of indoor pollution in the home and other buildings is of more recent origin. Consequently, the only federal legislation that directly considers indoor pollution in buildings focuses only on to radon, lead paint poisoning, and asbestos.

In addition to these specific laws, many environmental laws have indirect application to the problem of indoor pollution.

Major Indoor Pollution Laws

Radon

Radon Gas and Indoor Air Quality Research Act of 1986, Public Law 99-499, Title IV, October 17, 1986, 100 Statute 1758.

Congressional Investigation

Congress passed this act in response to findings that:

1. High levels of radon gas pose a serious health threat in structures in certain areas of the country;
2. Various scientific studies have suggested that exposure to radon,

including exposure to naturally occurring radon and indoor air pollutants, poses a public health risk;

3. Existing federal radon and indoor air pollutant research programs are fragmented and underfunded; and
4. An adequate information base concerning exposure to radon and indoor air pollutants should be developed by the appropriate federal agencies.

Research Program

This act requires that the Environmental Protection Agency establish a research program to consider the importance of not only radon gas but other pollutants of indoor air quality. This program is to be designed to:

1. Gather data and information on all aspects of indoor air quality in order to contribute to the understanding of health problems associated with the existence of air pollutants in the indoor environment;
2. Coordinate federal, state, local, and private research and development efforts relating to the improvement of indoor air quality; and
3. Assess appropriate federal government actions to mitigate the environmental and health risks associated with indoor air quality problems.

The program will include:

1. Research and development concerning the identification, characterization, and monitoring of the sources and levels of indoor air pollution, including radon. The program is to include:
   a. The measurement of various pollutant concentrations and their strengths and sources;
   b. The establishment of practices to determine high-risk building types; and
   c. Development of instruments for indoor air quality data collection.
2. Research relating to the effects of indoor air pollution and radon on human health;

3. Research and development relating to control technologies or other types of mitigation measures to prevent or abate indoor air pollution (including the development, evaluation, and testing of individual and generic control devices and systems);
4. Demonstration of methods for reducing or eliminating indoor air pollution and radon; including sealing, venting, and other methods that the EPA determine may be effective;
5. Research to be carried out in conjunction with the secretary of housing and urban development, for the purpose of developing:
    a. Methods for assessing the potential for radon contamination of new construction, including (but not limited to) consideration of the moisture content of soil, porosity, and radon content of soil; and
    b. Design measures to avoid indoor air pollution; and
6. The dissemination of information to assure the public availability of the findings of the research activities.

Asbestos

Asbestos School Hazard Detection and Control Act of 1980, Public Law 96-270, June 14, 1980, 94 Statute 487 to 498.

Congressional Investigation

In order to consider the need for legislation Congressional investigation found that:

1. Exposure to asbestos fibers has been identified after long study by reputable medical doctors, that asbestos significantly increases the incidence of lung cancer and other severe or fatal diseases;
2. Medical evidence has suggested that children may be particularly vulnerable to environmentally induced cancers;
3. Medical science has not established any minimum level of exposure to asbestos fibers which is considered to be safe to individuals exposed to the fibers;

4. Substantial amounts of asbestos, particularly in sprayed form, was used in school buildings, especially during the period 1946 through 1972;
5. Partial surveys in some states indicated that (a) in a number of school buildings asbestos fibers have become damaged or friable, causing asbestos fibers to be dislodged into the air, and (b) asbestos concentrations far exceed normal ambient air levels in school buildings containing damaged materials;
6. The presence in school buildings of friable or easily damaged asbestos creates an unwarranted hazard to schoolchildren's and school employees' health;
7. The Department of Health and Human Welfare and the Environmental Protection Agency, as well as other agencies, have attempted to advertise the health hazard, but no systematic program for identifying health hazards exists in schools;
8. Without an improved program of information distribution, technical and scientific assistance, and financial support, many local educational agencies and states will not be able to mitigate the potential asbestos school hazards; and
9. The effective regulation of interstate commerce for the protection of the public health requires the establishment of programs under this act to identify and mitigate asbestos hazards.

Purpose

As a response to the findings, the purpose of the act was to establish a program for the inspection of schools to detect the presence of hazardous asbestos materials, to provide loans to states or local educational agencies to contain or remove hazardous asbestos materials from schools, and to replace such materials with other suitable materials.

Task Force

This act requires the secretary of education to establish the Asbestos Hazards School Safety Task Force. The duties of the Task Force include:

1. Compilation of medical, scientific, and technical information;

2. Distribution of the information to all appropriate places;
3. Review of applications for grants and loans;
4. Review any guidelines established by the Environmental Protection Agency for identifying those schools in which exposure to asbestos fibers constitutes a health problem; and
5. Assist in the formulating of standards and procedures for investigating the asbestos health hazard.

Administrative Funding

Each state that received administrative funds for any applicable program is required to submit a report that:

1. Detailed the names in which the state distributed information to local educational agencies on the health hazards of asbestos; and
2. Described the procedures used by the state in maintaining records on asbestos detection, containment, or removal activities conducted by local educational agencies, and repairs made to restore school buildings to conditions comparable to those existing before the contaminant or removal activities occurred.

The act provides authority to the secretary of education to make grants to local educational agencies for the federal share of the costs of carrying out an asbestos detection program meeting federal standards.

An application for a grant had to:

1. Contain a description of the methods to be used by the local educational agency to determine whether hazardous concentrations of asbestos fibers or materials emitting such fibers exist in school buildings;
2. Contain an estimate of total cost of the detection program;
3. Designate the party which shall conduct the testing; and
4. Contain assurance that the program will be carried out in accordance with established federal standards.

The application loans for asbestos removal established under the secretary of education had to contain:

1. The nature of the asbestos problem;
2. The asbestos content of the material to be removed; and
3. The method to be used to remove the asbestos.

The application also had to contain assurance that:

1. The employee engaged in any asbestos removal activity had to be notified of the hazards of working with asbestos;
2. No child or school employee be permitted in the vicinity of any asbestos containment or removal activity; and
3. All asbestos removal employees will be paid reasonable rates of pay.

The secretary of education had to report:

1. The number of loans in the preceding year;
2. Each asbestos problem;
3. The types of programs in which loans were made;
4. Estimates of total costs; and
5. Number of loan applications disapproved.

Standards

The Task Force was required to establish standards and safety procedures that included:

1. Procedures for testing the level of asbestos fibers in schools;
2. Standards for evaluating the likelihood of the leakage of asbestos fibers into the school environment; and
3. Standards for determining which contractors are qualified to carry out the testing and evaluation of the asbestos problem.

The act also provided the means for the U.S. government to recover the cost of asbestos removal if a suit was brought against the original installer of the asbestos.

Asbestos School Hazard Abatement Act of 1984. Public Law 98-377, Title V, August 11, 1984, 98 Statute 1287.

Amended

In 1984, Congress amended the Asbestos School Hazard Detection and Control Act of 1980 to place the implementation of the program under the Environmental Protection Agency. It was now known as the Asbestos School Hazard Abatement Program.

Asbestos Hazard Emergency Response Act of 1986. Public Law 99-519. October 22, 1986, 100 Statute 2970–2991.

Amended

Public Law 100-368, July 18, 1988, 102 Statute 833

Public Law 101-637, November 28, 1990, 104 Statute 4590-4592, 4594

Public Law 103-382, Title III, Part I, October 20, 1994, 108 Statute 4027

Congressional Investigation

Congressional investigation found that additional legislation was required to reduce the asbestos hazard in the nation. The amendment to the 1980 act was placed under Title 22 of the Toxic Control Act of 1986. The 1986 Asbestos Hazard Emergency Response Act has been amended three times, in 1988, 1990, and 1994, but each contains the basic provisions of the original act.

The Congressional investigation of 1986 revealed the following:

1. It was found that the Environmental Protection Agency's regulations for local educational agency inspection for, and notification of, the presence of friable asbestos-containing material in school buildings included neither standards for the proper identification of asbestos-containing material or appropriate response actions with respect to friable asbestos-containing materials, nor a requirement that response actions with respect to friable asbestos-containing material be carried out in a safe and complete manner once actions were found to be

necessary. As a result of the lack of regulatory guidance from the Environmental Protection Agency, some schools have not undertaken the response action, whereas many others have undertaken expensive projects without knowing if such action was necessary, adequate, or safe. Thus, the danger of exposure to asbestos continues to exist in schools, and some exposure to asbestos continues to exist in schools; some exposure actually may have increased due to the lack of federal standards and improper response action;

2. There is no uniform program for accrediting persons involved in asbestos identification and abatement, nor are local educational agencies required to use accredited contractors for asbestos work;
3. The guidance provided by the Environmental Protection Agency in its "Guidelines for Controlling Asbestos-Containing Material in Buildings" is insufficient in detail to ensure adequate responses. Such guidance is intended to be used only until the regulations required by this title become effective; and
4. Because there are no federal standards whatsoever regulating daily exposure to asbestos in other public and commercial buildings, persons, in addition to those comprising the nation's school population, may be exposed daily to asbestos.

Purpose

As a result of the continuing lack of control of asbestos, the purpose of this act is:

1. To provide for the establishment of federal regulations that require inspection for asbestos-containing material and implementation of appropriate response action with respect to asbestos-containing material in the nation's schools in a safe and complete manner;
2. To mandate safe and complete periodic reinspection of school buildings following response actions, when appropriate; and
3. To require the Environmental Protection Agency to conduct a study to find out the extent of the danger to

human health posed by asbestos in public and commercial buildings and the means to respond to any such danger.

Asbestos Control

The regulations to control asbestos to be developed by the Environmental Protection Agency are to protect human health and the environment. The procedures include:

1. Inspection. An inspection for determining whether asbestos-containing material is present in a school building. The regulations provide for the exclusion of any school building if (a) inspection of such school building was completed before the effective date of the regulations, and (b) if that inspection meets the procedures and other requirements of the regulation in the "Guidelines for Controlling Asbestos-Containing Material in Buildings";
2. Circumstances Requiring Response Actions. The EPA regulations defined the appropriate response action for a school building controlled by a local educational agency:
   a. Damaged—Circumstances in which friable asbestos-containing material or its covering is significantly damaged, deteriorated, or delaminated;
   b. Significant Damage—Circumstances in which friable asbestos-containing material or its covering is significantly damaged, deteriorated, or delaminated;
   c. Potential Damage—Circumstances in which friable asbestos-containing material is in an area regularly used by building occupants, including maintenance personnel, in the course of their normal activities, or there is a good possibility that in the future the asbestos-containing material will be damaged, deteriorated, or delaminated; and
   d. Potential Significant Damage—An area where the potential for damage is great;
3. Response Action. The EPA is to develop regulations describing the types of response action for a school

building under a local educational agency, using the least burdensome methods, which protect human health and the environment. These regulations will take into account local circumstances including occupancy and use patterns and short- and long-term costs;

4. Implementation. The regulations of the EPA should include standards for the education and protection of both workers and building occupants for the following activities:
   a. Inspection;
   b. Response Action; and
   c. Post-Response Action, including periodic reinspections of asbestos-containing material and long-term surveillance of activities.
5. Operation and Maintenance. The EPA is required to prepare guidelines for the removal and maintenance of asbestos-free school buildings under the control of a local educational agency;
6. Periodic Surveillance. The EPA is to develop regulations for surveillance and periodic reinspection of friable and nonfriable asbestos in school buildings and the education of school employees about the location and safety procedures with respect to friable and nonfriable asbestos;
7. Transportation and Disposal. The EPA is to develop regulations that prescribe standards of transportation and disposal of asbestos-containing waste material to protect human health and the environment. These regulations will contain provisions as to how transportation vehicles are loaded and unloaded so as to assure the physical integrity of containers of asbestos-containing waste material; and
8. Management Plans. The EPA regulations require that each local educational agency develop an asbestos management plan for school buildings under its authority.

The management is to include:

   a. An inspection statement describing inspection and response action activities;

b. A description of the results of the inspection conducted pursuant to regulations;
c. A detailed description of measures to be taken to respond to any friable asbestos-containing material;
d. A detailed description of any asbestos-containing material that remains in the school building once response action is undertaken;
e. A plan for periodic reinspection and long-term surveillance activities;
f. A statement as to whether the state has adopted a contractor accreditation plan, or if the local agency uses contractors accredited in another state;
g. A list of laboratories that analyze bulk samples of asbestos-containing materials;
h. A list of consultants who contributed to the management plan;
i. An evaluation of the resources needed to successfully complete response actions and carry out reinspection, surveillance, and operation and maintenance activities;
j. A statement from the contractor that he is able to comply with the management plan;
k. The EPA regulations require that each local educational agency attach a warning label to any asbestos-containing material still in maintenance areas. The warning label is to say: Caution: Asbestos. Hazardous. Do Not Disturb Without Proper Training and Equipment;
l. The management plan may be submitted in stages;
m. A copy of the management plan is to be made public; and
n. The management plan may be changed as the project proceeds.

The act provides procedures for the local educational agencies to develop local plans for the removal of asbestos if the EPA fails to provide regulations.

Accreditation of Contractors

The act provides guidelines for the accreditation of contractors. It states that the EPA will develop a model contractor accreditation plan for states to give accreditation to persons in the following categories:

1. Persons who inspect for asbestos-containing material in school buildings under the authority of a local educational agency;
2. Persons who prepare management plans for such schools; and
3. Persons who design or carry out response actions.

Model for Control

The accreditation model shall include a requirement that all persons must achieve a passing grade on an examination and participate in continuing education to stay informed about current asbestos inspection and response action technology. The examination may include requirements for knowledge in the following areas:

1. Recognition of asbestos-containing material and its physical characteristics;
2. Health hazards of asbestos and the relationship between asbestos exposure and disease;
3. Assessing the risk of asbestos exposure through a knowledge of percentage weight of asbestos-containing material, friability, age, deterioration, location, and accessibility of materials and advantages and disadvantages of dry and wet response action methods;
4. Respirators and their use, including case, selection, degree of protection afforded, testing, and maintenance and cleaning procedures;
5. Appropriate work practices and control methods, including the use of high-efficiency particle vacuums, the use of water, and principles of negative air pressure equipment use and procedures;
6. Preparing a work area for response action work, including isolated work areas to prevent public exposure to asbestos, decontamination procedures, and procedures for dismantling work areas after completion of work;
7. Establishing emergency procedures to respond to sudden release of asbestos;
8. Air monitoring requirements and procedures;
9. Medical surveillance program requirements;
10. Proper asbestos waste transportation and disposal procedures; and

11. Housekeeping and personal hygiene practices, including the necessity of showers, and procedures to prevent asbestos exposure to an employee's family.

Accreditation Requirements

Under the requirements of the act a person may not:

1. Inspect for asbestos-containing material in a school building under the authority of a local educational agency;
2. Prepare a management plan for such a school; or
3. Design or conduct response actions with respect to friable asbestos-containing material in such a school unless that person is accredited by a state program or an EPA-approved course.

The act also provides for the development of an accreditation program for laboratories by the National Bureau of Standards. The National Bureau of Standards is to: develop an accreditation program for two types of laboratories—those that conduct qualitative and semiquantitative analyses of both samples of asbestos-containing material and those that conduct analyses of air samples of asbestos from school buildings under the authority of a local educational agency.

Enforcement of Legislation

The act provides policies for enforcement of the law. Any civil penalty collected is to be used to comply with the law. There is also a provision that citizens may file a complaint with the EPA or the governor of the state. In addition, any person may petition the EPA to initiate a proceeding for the issuance, amendment, or repeal of a regulation or order under this act. The petition must provide facts as to why it is necessary to issue, amend, or repeal a regulation. Public hearings may be held on the petition requested. The EPA will either grant or deny the petition. If approved, the EPA will implement the request. If denied, the EPA will publish in the *Federal Register* the reason for the denial.

Emergency Planning

Emergency action is also provided for in the act. This authority is invoked by either the EPA or the governor of the state when:

1. The presence of airborne asbestos or the condition of friable asbestos-containing material in a school building controlled by a local educational agency poses an imminent and substantial endangerment to human health or the environment; and
2. The local educational agency is not taking sufficient action (as determined by EPA or the governor) to respond to the airborne asbestos or friable asbestos-containing material.

In order to implement the emergency provision the following notification must occur:

1. In the case of the governor taking action, the governor shall notify the local educational agency concerned; and
2. In the case of the EPA taking action, the EPA shall notify the local educational agency concerned and the governor of the state in which such agency is located.

Injunctive Relief

In order to bring an action for injunctive relief from the presence of airborne asbestos or asbestos-containing material, it may be necessary for:

1. The EPA to request the Attorney General to bring suit; or
2. The governor of the state to bring suit.

The district court of the United States in the district in which the response will be carried out shall have jurisdiction to grant such relief, including injunctive relief.

Liability Insurance

The act provides guidelines on the availability of liability insurance and other forms of assurance against financial loss that are available to local educational agencies and asbestos contractors with respect to actions required to remove asbestos.

Public protection is also provided in that no state or local educational agency may discriminate against a person, including the firing of an employee, because the person provided information relating to a potential violation of the act to any other person, including the state and federal governments. A person who has been fired or discriminated against may appeal, within 90 days, to the secretary of labor for review of the circumstances.

The EPA is required by the act to appoint an ombudsman whose duties are:

1. To receive complaints, grievances, and requests for information submitted by any person;
2. To render assistance with respect to the complaints, grievances, and requests received; and
3. To make such recommendations to the EPA as an ombudsman considers appropriate.

Congress Report

The EPA is required to submit to Congress the results of a study that shall:

1. Assess the extent to which asbestos-containing materials are present in public and commercial buildings;
2. Assess the condition of asbestos-containing materials in commercial buildings and the likelihood that persons occupying such buildings, including service and maintenance personnel, are, or may be, exposed to asbestos fibers;
3. Consider and report on whether public commercial buildings should be subject to the same inspection and response action requirements that apply to school buildings;
4. Assess whether existing federal regulations adequately protect the general public, particularly abatement procedures, from exposure to asbestos during renovation and demolition of such buildings; and
5. Include recommendations that explicitly address whether there is a need to establish standards in order to regulate asbestos exposure in public and commercial buildings.

Asbestos School Hazard Abatement Reauthorization Act of 1990, Public Law 101-637, November 28, 1990, 104 Statute 4589.

Environmental Protection Agency Studies

Although initial legislation had been enacted to remove asbestos from buildings in 1984, more recent studies by the EPA revealed that problems of asbestos contamination persist. The EPA report to Congress revealed the following:

1. The Environmental Protection Agency estimated that more than 44,000 school buildings contain friable asbestos, exposing more than 15 million schoolchildren and 1.5 million school employees to unwarranted health hazards;
2. To remove asbestos from buildings the EPA estimated it will cost local educational agencies more than $3 billion to comply with the Asbestos Hazard Emergency Response Act;
3. Without a continuing program of information, technical and scientific assistance, training, and financial support, many local educational agencies will be unable to carry out sufficient response actions to prevent the release of asbestos fibers into the air;
4. Without the provisions of sufficient financial support, the cost to local educational agencies of implementing asbestos response actions may have an adverse impact on their educational mission; and
5. Effective regulations of asbestos for the protection of human health and the environment requires the continuation of programs to mitigate the hazards of asbestos fibers.

Purpose

The purpose of this act is to reaffirm the previous asbestos legislation and to:

1. Direct the EPA to maintain a program to assist local schools in carrying out their responsibilities under the Asbestos Hazard Emergency Response Act;
2. Provide continuing scientific and technical assistance to state and local agencies to enable them to identify and abate asbestos health hazards;
3. Provide financial assistance to state and local agencies for training of persons involved with inspections and

abatement of asbestos, for conducting necessary reinspections of school buildings, and for the actual abatement of asbestos threats to the health and safety of schoolchildren or employees; and

4. Assure that no employee of a local educational agency suffers any disciplinary action as a result of calling attention to potential asbestos hazards that may exist in schools.

Asbestos Information Act of 1988, Public Law 100-577, October 31, 1988. 102 Statute 2901.

This act requires that the Environmental Protection Agency provide the public with additional information about asbestos products. In order to do this the act stipulated that prior to 1988 any manufactured or processed asbestos or asbestos-containing material that was prepared for sale for use as surfacing material, thermal system insulation, or miscellaneous material in buildings shall submit to the EPA the year of manufacture, the types or classes of products, and to the extent available, other identifying characteristics reasonably necessary to identify or distinguish the asbestos or asbestos-containing material.

Lead Poisoning

Lead-Based Paint Poisoning Prevention Act of 1971, Public Law 91-695, January 13, 1971, 84 Statute 2078.

Amended

Public Law 93-151, November 9, 1973, 87 Statute 565 to 567

Public Law 94-317, Title II, June 23, 1976, 90 Statute 705, 706

Public Law 100-242, Title V, February 5, 1988, 101 Statute 1945

Public Law 100-628, Title X, November 7, 1988, 102 Statute 3280 to 3282

Public Law 102-550, Title X, October 28, 1992, 106 Statute 3904, 3907

Investigation of Lead Poisoning

This act provides federal assistance to help develop and implement local programs to eliminate the causes of lead-based paint poisoning, to detect and treat incidences of such poisoning, to establish a federal demonstration and research program to study the extent of the lead-based paint poisoning problem and methods available for lead-based paint removal, and to prohibit future use of lead-based paint in federal or federally assisted construction or rehabilitation.

Reduction and Treatment

Title I of the bill provides the means for detection and treatment of lead-based paint poisoning. The act requires that the secretary of health, education, and welfare provide grants not exceeding 75 percent of the cost of developing and carrying out the local program. The local program is to include:

1. Educational programs intended to communicate the health danger and prevalence of lead-based paint poisoning to children of inner-city areas, parents, educators, and local health officials;
2. Develop and carry out intense community testing programs designed to detect incidents of lead-based paint poisoning among community residents and to ensure prompt medical treatment for such affected individuals;
3. Develop and implement intensive follow-up programs to ensure that identified cases of lead-based paint poisoning are protected against further exposure to lead-based paints in their living environment; and
4. Any other actions necessary to reduce lead poisoning.

Program

Title II of the act aided the development of programs that identify areas of high risk to the health of residents because of the presence of lead-based paints on interior surfaces. To assist in the removal of lead-based paint, grant proposals were available.

Grant Proposal

The proposal for a grant must include:

1. Development of the means to carry out a comprehensive testing program to detect the presence

of lead-based paints on surfaces of residential housing; and

2. Development and implementation of a comprehensive program requiring the prompt elimination of lead-based paints from all interior surfaces, porches, and exterior surfaces to which children might commonly be exposed, from residential housing on which lead-based paints have been used as a surface covering, including those surfaces on which non–lead-based paints have been used to cover surfaces to which lead-based paints were previously applied.

Research Program

Title III of the act requires the secretary of housing and urban development with the secretary of health, education, and welfare to develop a demonstration and research program to determine the nature and extent of the problem of lead poisoning in the United States, particularly in urban areas, and the methods by which lead-based paint can most effectively be removed from interior surfaces, porches, and exterior surfaces to which children are commonly exposed in residential housing.

Title IV of the act indicates that the secretary of health, education, and welfare shall develop measures to prohibit the use of lead-based paint in residential structures.

Lead Contamination Control Act of 1988. Public Law 100-572, October 31, 1988, 102 Statute 2884.

This act amends the Safe Drinking Water Act of 1974 (Public Law 95-523, December 16, 1974, 88 Statute 1660). The 1974 act was enacted to assure that the public was provided with safe drinking water.

Control of Lead

The amended Safe Drinking Water Act was specifically designed to control lead in drinking water. The implementation was through the control of watercoolers. A watercooler was defined as any mechanical device affixed to drinking water supply plumbing that cools water for human consumption. To have lead-free water from a cooler the following conditions had to exist:

1. No part or component of these coolers that contains more than 8 percent lead can come into contact with the drinking water;
2. Any cooler than contains any solder, flux, or storage tank interior surface that may come into contact with drinking water shall be considered lead-free if the solder, flux, or storage tank interior surface contains no more than 0.2 percent lead; and
3. More stringent requirements may be imposed if any part of the watercooler constitutes an important source of lead in drinking water.

Content of Watercoolers

The act provides that a list of brands and models of watercoolers that are not lead-free be published based on the best information available from the Environmental Protection Agency. All water-coolers sold in interstate commerce had to be lead-free.

The act directs particular attention to lead contamination in schools' drinking water. Guidelines were to be published to determine the source and degree of lead contamination in schools' drinking water supplies and in remedying such contamination. The Remedial Action Program included:

1. Within nine months of enactment of this act each state was to establish a program to assist local educational agencies in remedying lead contamination in drinking water from coolers and other sources of lead contamination in schools;
2. Publish availability of all procedures to eliminate lead contamination;
3. Within 15 months of this act all school coolers will be removed or repaired so that they were free of lead contamination.

Federal aid was provided in the act to implement the program.

Public Health Service Act, Public Law 100-471, October 4, 1988, 102 Statute 2284.

In 1988 the Public Health Service Act was amended to add a new section, 317A, Lead Poisoning Prevention. This section is known as the Lead Contamination Control Act of 1988, Public Law 100-572, October 31, 1988, 102 Statute 2784.

This act provides that the secretary of health, education, and welfare, acting through the Director of the Centers for Disease Control, will provide grants to states and agencies of local governments for the initiation and expansion of community programs to:

1. Screen infants and children for elevated blood-lead levels;
2. Assure referrals for treatment of infants and children with elevated blood-lead levels; and
3. Provide education about childhood lead poisoning.

The grant application must include:

1. A complete description of the program; and
2. Assurance that reports will be made quarterly on the number of infants and children screened for elevated blood-lead levels, the number with elevated blood-lead levels, and the number referred for treatment.

The act also provides the requirement that the administrator of the Environmental Protection Agency assure that laboratories that test drinking water supplies for lead contamination be certified to provide reliable, accurate testing. The program also requires EPA to develop educational programs designed to communicate to parents, educators, and local health officials the significance and prevalence of lead poisoning in infants and children.

Residential Lead-Based Paint Hazard Reduction Act of 1992, Public Law 102-550, Title X, October 28, 1992, 106 Statute 3897 to 3911, and 3924 to 3927.

This is the most comprehensive act passed by Congress to control and eliminate the hazards of lead-based paint. Although the problem was recognized by the 1971 act of Congress, the lead-based paint problem continued to persist.

Congress Investigation

After investigating the problem, Congress found that:

1. Low level lead poisoning continued to be widespread among American children, affecting as

many as 3 million under the age of six, with minority and low-income communities disproportionately affected;

2. At low levels, lead poisoning in children causes intelligence quotient deficiencies, reading and learning disabilities, impaired hearing, reduced attention span, hyperactivity, and behavior problems;
3. Pre-1980 U.S. housing contains more than 3 million tons of lead in the form of lead-based paint;
4. The ingestion of household dust containing lead from deteriorating or abraded lead-based paint is the most common cause of lead poisoning in children;
5. The health and development of children living in as many as 3.8 million U.S. homes are endangered by chipping or peeling lead paint, or excessive amounts of lead-contaminated dust;
6. The dangers posed by lead-based paint can be reduced by abating lead-based paint or by taking interim measures to prevent paint deterioration and limit children's exposure to lead dust and chips;
7. Despite the enactment of laws in the early 1970s requiring the federal government to eliminate as far as practicable lead-based paint hazards in federally owned, assisted, and insured housing, the federal response to this national crisis remains severely limited; and
8. The federal government must take a leadership role in building the infrastructure—including an informed public, certified inspectors and contractors, laboratories, trained workers, available financing and insurance—necessary to ensure the national goal of eliminating lead-based paint hazards in housing can be achieved as expeditiously as possible.

Purpose

In response to these findings the purposes of this new act were:

1. To develop a national strategy to build the infrastructure necessary to eliminate lead-based paint hazards in all housing as expeditiously as possible;

2. To reorient the national approach to the presence of lead-based paint in housing and to implement, on a priority basis, a broad program to evaluate and reduce lead-based paint hazards in the nation's housing;
3. To encourage effective action to prevent childhood lead poisoning by establishing a workable framework for lead-based paint hazard evaluation and reduction, and by ending the current confusion over reasonable standards of care;
4. To ensure that the existence of lead-based paint hazards is taken into account in the development of government housing policies and in the sale, rental, and renovation of homes and apartments;
5. To mobilize national resources expeditiously, through a partnership among all levels of government and the private sector, to develop the most promising, cost-effective methods for evaluating and reducing lead-based paint hazards;
6. To reduce the threat of childhood lead poisoning in housing owned, assisted, or transferred by the federal government; and
7. To educate the public concerning the hazards and sources of lead-based paint poisoning and steps to reduce and eliminate such hazards.

Lead Evaluation

This act authorizes the secretary of housing and urban development to provide grants to eligible applicants to evaluate and reduce lead-based paint hazards in priority housing that is not federally assisted, federally owned, or public housing.

The selection criteria includes:

1. The extent to which the proposed activities will reduce the risk of lead-based paint poisoning to children under the age of six who reside in priority housing;
2. The degree of severity and extent of lead-based paint hazards;
3. The ability of the applicant to secure additional funds from state, local, or private sources to supplement the federal grant; and

4. The ability of the applicant to carry out the proposal activities.

Federal Grant

A federal grant may be used to carry out eligible activities:

1. Perform risk assessments and inspections in priority housing;
2. Provide for the interim control of lead-based paint hazards in priority housing;
3. Provide for the abatement of lead-based paint hazards in priority housing;
4. Provide for the additional cost of reducing lead-based paint hazards in units undergoing renovation funded by other sources;
5. Ensure that risk assessments, inspections, and abatements are carried out by certified contractors;
6. Monitor the blood-lead levels of workers involved in lead hazard reduction activities;
7. Assist in the temporary relocation of families forced to vacate priority housing while lead hazard reduction measures are being conducted;
8. Educate the public on the nature and causes of lead poisoning and measures to reduce exposure to lead, including exposure due to residential lead-based paint hazards; and
9. Test soil and interior surface dust, levels of buildings, and the blood-lead levels of children under the age of six residing in priority housing after lead-based paint hazard reduction activity has been conducted, and to assure that such activity does not cause excessive exposures to lead.

Information on Lead Poisoning

This act also provides for the disclosure of information concerning lead upon transfer of residential property. Within two years of the enactment of this act, the secretary of welfare and urban development and the administrator of the Environmental Protection Agency shall develop regulations that disclose the lead-based paint hazards in buildings that are offered for sale or lease. The seller or lease holder must:

1. Provide the purchaser or lessee with a lead hazard information report;
2. Disclose to the purchaser or lessee the presence of any known lead-based paint, or any known lead-based hazards, in such housing and provide to the purchaser or lessee any lead hazard evaluation report available; and
3. Permit the purchaser a ten-day period to conduct risk assessment or inspection for the presence of lead-based paint hazards.

Lead Poisoning Warning

In addition to the above regulations, the act requires that every contract for the purchase or sale of target housing contain a lead warning statement that the purchaser has:

1. Read the Lead Warning Statement and understands the contents;
2. Received a lead hazard information pamphlet; and
3. Had a ten-day opportunity to conduct a risk assessment or inspection for the presence of lead-based paint hazards.

The law provides that any person who knowingly violates the disclosures of information of the hazards of lead poisoning is subject to civil penalties equal to three times the amount of damages incurred by the purchase of the housing.

Worker Protection

This act also protects workers by regulating occupational exposure to lead in the construction industry. To implement the act, the secretary of housing and urban development, in cooperation with other federal agencies, is to conduct research on strategies to reduce the risk of lead exposure including interior lead dust in carpets, furniture, and forced airducts. The research will:

1. Develop improved methods for evaluating lead-based paint hazards in housing;
2. Develop improved methods for reducing lead-based paint hazards in housing;
3. Develop improved methods for measuring lead in paint, films, dust, and soil samples;

4. Establish performance standards for various detection methods, including spot test kits;
5. Establish performance standards for lead-based paint hazard reduction methods, including the use of encapsulants;
6. Establish appropriate clean-up standards;
7. Evaluate the efficacy of interim controls in various hazardous situations;
8. Evaluate the relative performance of various abatement techniques;
9. Evaluate the long-term cost-effectiveness of interim control and abatement strategies; and
10. Assess the effectiveness of hazard evaluation and reduction activities funded by this act.

The comptroller general of the United States is to assess the effectiveness of federal enforcement and compliance with lead safety laws and regulations and is also to assess the availability of liability insurance for owners of residential housing that contains lead-based paint and persons engaged in lead-based paint hazard evaluation and reduction activities.

The assessment is to include:

1. An analysis of any precedents in the insurance industry for the containment and abatement of environmental hazards, such as asbestos, in federally assisted housing;
2. An assessment of the recent insurance experience in the public housing lead hazard identification and reduction program; and
3. Measures for increasing the availability of liability insurance to owners and contractors engaged in federally supported work.

Each year the secretary of housing and urban development must report to Congress on the progress of implementing the provisions of the act.

Lead-Based Paint Exposure Reduction Act of 1992, Public Law 102-550, Title X, Subtitle B, October 28, 1992, 106 Statute 3912 to 3924, is part of the Toxic Substance Control Act of 1992, which is amended by adding after Title III the following new title: Title IV Lead Exposure Reduction.

Regulation Development

This act requires that regulations be developed and implemented governing lead-based paint activities to ensure that individuals engaged in such activities are properly trained, that training programs are accredited, and that contractors are certified. Such regulations are to contain standards for performing lead-based paint activities, taking into account reliability, effectiveness, and safety. Such regulations shall require that all risk assessment, inspection, and abatement activities performed in target housing shall be performed by certified contractors.

Final regulations are to contain specific requirements for the accreditation of lead-based paint activities training programs for workers, supervisors, inspectors, planners, and other individuals engaged in lead-based paint activities, including:

1. Minimum requirements for the accreditation of training providers;
2. Minimum training curriculum requirements;
3. Minimum training hour requirements;
4. Minimum hands-on training requirements;
5. Minimum training competency and proficiency requirements; and
6. Minimum requirements for training program quality control.

Programs

The act authorizes the establishment of state programs. The approval of these programs is to be based on:

1. The necessity to protect human health; and
2. Adequate enforcement.

In order to aid states in the development of a program, the act requires that a model state program, which may be adopted by the states, be proposed. The model program is to encourage states to utilize existing state and local certification and accreditation programs and procedures.

Comprehensive Program

The act also requires that the Environmental Protection Agency conduct a comprehensive program to promote safe, effective, and affordable monitoring, detection, and abatement of lead-based paint and other lead exposure hazards. In the process, EPA is to establish protocols, criteria, and minimum performance standards for laboratory analysis of lead in paint films, soil, and dust. Within two years of the passage of the act, a program was to be established to certify laboratories qualified to test substances for lead content unless voluntary accreditation programs are in place and operating on a nationwide standard.

The secretary of health and human development with consultation from the director of the National Institute for Occupational Safety and Health is to conduct a comprehensive study of means to reduce hazardous occupational lead abatement exposure. This study will include:

1. Surveillance and intervention capability in the states to identify and prevent hazardous exposures to lead abatement workers;
2. Demonstration of lead abatement control methods and devices and work practices to identify and prevent hazardous lead exposures in the workplace;
3. Evaluation, in consultation with the National Institute of Environmental Health Sciences, of health effects of low and high levels of occupational lead exposures on reproductive, neurological, renal, and cardiovascular health;
4. Identification of high risk occupational settings to which prevention activities and resources should be targeted; and
5. A study assessing the potential exposures and risks from lead to janitorial and custodial workers.

Sources of Lead Poisoning

These studies are to examine relative contributions to elevated lead body burden from each of the following:

1. Drinking water;
2. Food;
3. Lead-based paint and dust from lead-based paint;
4. Exterior sources such as ambient air and lead in soil;
5. Occupational exposure; and
6. Other exposures.

Public Education

The act provides for public education and outreach activities to increase public awareness of:

1. The scope and severity of lead poisoning from household sources;
2. Potential exposure to sources of lead in schools and childhood daycare centers;
3. The implications of exposures for men and women, particularly women of childbearing age;
4. The need for quality abatement and management actions;
5. The need for universal screening of children;
6. Other components of a lead poisoning prevention program;
7. The health consequences of lead exposure resulting from lead-based paint hazards;
8. Risk assessment and inspection methods for lead-based paint hazards; and
9. Measures to reduce the risk of lead exposure from lead-based paint.

These activities are to provide educational services and information to:

1. Health professionals;
2. The general public, with emphasis on parents of young children;
3. Homeowners, landlords, and tenants;
4. Consumers of home improvement products;
5. The residential real estate industry; and
6. The home renovation industry.

The act also provides for the establishment of a National Clearinghouse on Childhood Lead Poisoning. The clearinghouse will:

1. Collect, evaluate, and disseminate current information on the assessment and reduction of lead-based paint hazards and advise on health effects, sources of exposure, detection and risk assessment methods,

environmental hazards abatement, and clean-up standards; and

2. Maintain a rapid-alert system to inform certified lead-based paint contractors of significant developments in research related to lead-based paint hazards.

A single lead-based paint hazard hotline is to be established to provide the public with answers to questions about lead poisoning prevention and referrals to the clearinghouse for technical information.

The Environmental Protection Agency, in coordination with the secretaries of housing and urban development and health and human services, is to prepare a pamphlet. The pamphlet is to:

1. Contain information regarding the health risks associated with exposure to lead;
2. Provide information on the presence of lead-based paint hazards in federally assisted, federally owned, and target housing;
3. Describe the risks of lead exposure for children under six years of age, pregnant women, women of childbearing age, persons involved in home renovation, and others residing in a dwelling with lead-based paint hazards;
4. Describe the risks of renovation in a dwelling with lead-based paint hazards;
5. Provide information on approved methods for evaluating and reducing lead-based paint hazards and their effectiveness in identifying, reducing, eliminating, or preventing exposure to lead-based paint hazards;
6. Advise persons how to obtain a list of contractors certified in lead-based paint hazard evaluation and reduction;
7. State that a risk assessment or inspection for lead-based paint is recommended prior to the purchase, lease, or renovation of target housing;
8. State that certain state and local laws impose additional requirements related to lead-based paint in housing and provide a listing of federal, state, and local agencies in each state, including addresses and telephone numbers that can provide information about

applicable laws and available governmental and private assistance and financing; and
9. Any other pertinent information.

This pamphlet must be given to anyone who performs renovation of target housing. The act also requires that all federal housing activities comply with the law.

Related Laws and Regulations

Tobacco

Public Health Cigarette Smoking Act of 1970, Public Law 91-222, April 1, 1970, 84 Statute 87.

By this act Congress established a comprehensive federal program to deal with cigarette labeling and advertising with respect to any relationship between smoking and health.

Federal Policy

The policy was:

1. The public was to be adequately informed that cigarette smoking may be hazardous to health by an inclusion of a warning to the effect on each package of cigarettes; and
2. Commerce and the national economy is:
   a. To be protected to the maximum extent consistent with this declared policy;
   b. Not to be impeded by diverse, nonuniform, and confusing cigarette labeling and advertising regulations with respect to any relationships between smoking and health.

Implementation

The act states that any person who manufactured, imported, or packaged cigarettes had to label the package with the statement: Warning: The Surgeon General Has Determined That Smoking is Dangerous to Your Health. The statement is to be located in a conspicuous place on every cigarette package and is to appear in conspicuous and legible type in contrast to typography, layout, and color with other printed matter on the package.

To further control the smoking of cigarettes, the act states that after January 1, 1971, it is unlawful to advertise cigarettes on any medium of electric communication subject to the Federal Communications Commission.

The law also provided for a preemption in that no statement related to smoking and health, other than the specified statement of the act, shall be required on any cigarette package, and no requirement or prohibition based on smoking and health is to be imposed under state law with respect to the advertising or promotion of any cigarettes, the package of which is labeled in conformity with the provision of this act.

The Federal Trade Commission has the responsibility of implementing this act. If the Federal Trade Commission determines it is necessary to take action with respect to violations of the act, it must notify Congress. Such text is to include the trade regulation rule and a full statement of the violation. Congress may act if it desires.

Any person found guilty of violating the provisions of this act shall be guilty of a misdemeanor and may be subject to a fine of not more than $10,000.

The provisions of this act do not apply to packages of cigarettes manufactured, imported, or packaged for export from the United States or for delivery to a vessel or aircraft, as supplies, for consumption beyond the jurisdiction of the internal revenue laws of the United States. This exemption does not apply to cigarettes manufactured, imported, or packaged for sale or distribution to members or units of the Armed Forces of the United States located outside the United States.

Energy Conservation—Sick Building Syndrome

The passage of the energy conservation acts after the energy crisis of the 1970s had the basic purpose of reducing energy consumption. One of the major goals was to reduce energy consumption in buildings by greater insulation. The endeavor did not recognize that as a response to "tighter" houses there was little change of air from the outside to the inside. As a consequence, pollution of the indoor air increased, causing a number of health problems. This situation has come to be known as sick building syndrome.

National Energy Conservation Policy Act of 1978, Public Law 95-619, November 9, 1978, 92 Statute 3206.

Amended

Public Law 96-294, Title IV, Title V, June 30, 1980, 716, 741-746, 752

Public Law 99-272, Title VII, April 7, 1986, 100 Statute 142

Public Law 99-412, Title I, August 28, 1986, 100 Statute 932-943

Public Law 99-509, Title III, October 21, 1986, 100 Statute 1890

Public Law 100-615, November 5, 1988, 102 Statute 3185-3189

Public Law 101-218, December 11, 1989, 103 Statute 1868

Public Law 102-54, June 13, 1991, 105 Statute 281

Public Law 102-486, Title I, Subtitle A, Subtitle F, October 24, 1992, 106 Statute 2787, 2844-2862

Public Law 104-66, Title I, Subtitle E, December 21, 1995, 109 Statute 718

Congressional Investigation

This act was a response to the energy crisis of the 1970s. It was enacted due to congressional findings that:

1. The United States faces an energy shortage arising from increasing demands for energy, particularly from insufficient domestic supplies of oil and natural gas;
2. Unless effective measures are promptly taken by the federal government and other users of energy to reduce the rate of growth of demand for energy, the United States will become increasingly dependent on the world oil market, increasingly vulnerable to interruption of foreign oil supplies, and unable to provide the energy to meet future needs; and
3. All sectors of our national economy must begin immediately to significantly reduce the demand for nonrenewable energy resources such as oil and natural

gas by implementing and maintaining effective conservation measures for the efficient use of these and other energy sources.

Purpose

The purposes of the act are to:

1. Provide for the regulation of interstate commerce;
2. To reduce the growth in demand for energy in the United States; and
3. To conserve nonrenewable energy resources produced in the nation and elsewhere without inhibiting beneficial economic growth.

Title II of the act, Residential Energy Conservation, and Title III, Energy Conservation Programs for Schools and Hospitals and Buildings Owned by Units of Local Governments and Public Care Institutions, provide regulations for the conservation of energy in buildings.

Residential Energy Conservation

This act requires the secretary of energy to prepare rules to include standards:

1. Necessary to assure general safety and effectiveness of any residential energy conservation measures;
2. Necessary for installation of any residential energy conservation measure;
3. For the procedures concerning fair and reasonable prices and rates of interest required;
4. Developed in consultation with the Federal Trade Commission concerning unfair, deceptive, or anticompetitive acts or practices;
5. For suppliers and contractors; and
6. Which assure that any person who alleges any injury resulting from a violation of the requirements of the standards shall be entitled to redress under provisions established by the governors of states.

After the secretary of energy established the building standards, each governor of the state or state agency had to submit a proposal as to how the state was to meet the proposed energy conservation plan.

Energy Conservation

The act also requires the secretary of energy to prepare a report on the potential for energy conservation in apartment buildings. The report had to include a consideration of:

1. Structural and energy control measures that may result in energy conservation in apartment buildings;
2. Potential for energy conservation in apartment buildings that could be achieved by the application of a utility program to apartment buildings;
3. The cost of achieving energy conservation in apartment buildings and the need for federal financial assistance to achieve energy savings; and
4. Recommendations for appropriate legislation.

Energy Conservation Standards for New Buildings Act of 1976, Public Law 94-385, Title III, August 14, 1976, 90 Statute 1144-1149.

Amended

Public Law 95-91, Title VII, August 4, 1977, 91 Statute 608

Public Law 95-619, Title II, November 9, 1978, 92 Statute 3238

Public Law 96-399, Title III, October 8, 1980, 94 Statute 3228

Public Law 97-35, Title X, August 13, 1981, 95 Statute 621

Public Law 100-242, Title V, February 5, 1988, 101 Statute 1950

Congressional Investigation

This act was based on congressional findings that:

1. Large amounts of fuel and energy are consumed unnecessarily each year in heating, cooling, ventilating, and providing domestic hot water for newly constructed residential and commercial buildings because such buildings lack adequate energy conservation features;

2. Federal performance standards for newly constructed buildings can prevent such waste of energy, which the nation no longer can afford in view of the current and anticipated energy shortages;
3. The failure to provide adequate energy conservation measures in newly constructed buildings increases long-term operating costs that may affect adversely the repayment of, and security for, loans made, insured, or guaranteed by federal agencies or made by federally insured or regulated instrumentalities; and
4. State and local building codes or similar controls can provide an existing means by which to assure, in coordination with other building requirements and with a minimum of federal interference in state and local transactions, that newly constructed buildings contain adequate energy conservation features.

Purpose

The purposes of the act are to:

1. Redirect federal policies and practices to assure that reasonable energy conservation features will be incorporated into new commercial and residential buildings receiving federal financial assistance;
2. Provide for the development and implementation of performance standards for new residential and commercial buildings that are designed to achieve the maximum practicable improvements in energy efficiency and increase in the use of nondepletable sources of energy; and
3. Encourage states and local governments to adopt and enforce such standards through their existing building codes and other construction control mechanisms, or to apply them through a special approval process.

Building Standards

This act requires that a new set of building standards be prepared for new commercial and residential buildings. The standards are to be published in the *Federal Register* and then finalized after public comment. Periodically the standards are to be brought up to date.

In order for states to receive federal assistance, they must be certified according to the standard regulations. In order to be certified, the state must indicate that:

1. The local government that has jurisdiction over the standards has adopted and is implementing a building code or other construction controls, which meet or exceed the requirements of approved performance standards; or
2. Each state has adopted and is implementing, on a statewide basis, a building code or laws or regulations that provide for the effective application of such final performance standards.

Control of Programs

The act also provides that the secretary of energy, with the assistance of the National Institute of Building Services, shall:

1. Monitor the progress made by the states and their political subdivisions in adopting and enforcing energy conservation standards in new buildings;
2. Identify any procedural obstacles or technical constraints inhibiting implementation of such standards;
3. Evaluate the effectiveness of such prevailing standards; and
4. Report to Congress the progress of the states and units of general progress of local government in adopting and implementing energy conservation standards for new buildings and the effectiveness of such standards.

Energy Conservation in Existing Buildings Act of 1976, Public Law 94-385, Title IV, August 14, 1976, 90 Statute 1150–1169.

Amended

Public Law 95-619, Title II, November 9, 1978, 92 Statute 3224

Public Law 96-294, Title V, June 30, 1980, 94 Statute 759

Public Law 98-181, Title IV, November 30, 1983, 97 Statute 1235

Public Law 98-479, Title II, October 17, 1984, 98 Statute 2228

Public Law 98-558, Title IV, October 30, 1984, 98 Statute 2887

Public Law 100-242, Title V, February 5, 1988, 101 Statute 1950

Congressional Investigation

This act was enacted by Congress in response to findings that:

1. The fastest, most cost-effective, and most environmentally sound way to prevent future energy shortages in the United States, while reducing the nation's dependence on imported energy supplies, is to encourage and facilitate, through major programs, the implementation of energy conservation and renewable-resource energy measures with respect to dwelling units, nonresidential buildings, and industrial plants;
2. Current efforts to encourage and facilitate such measures are inadequate as a consequence of:
   a. A lack of adequate and available financing for such measures, particularly with respect to individual consumers and owners of small businesses;
   b. A shortage of reliable and impartial information and advisory services pertaining to practicable energy conservation measures and renewable-resource energy measures and the cost savings that are likely if they are implemented in such units, buildings, and plants; and
   c. The absence of organized programs that, if they existed, would enable consumers, especially individuals and owners of small businesses, to undertake such measures easily and with confidence in their economic value;
3. Major financial incentives and assistance for energy conservation measures and renewable-resource energy measures in dwelling units, nonresidential buildings, and industrial plants would:

   a. Significantly reduce the nation's demand for energy and the need for petroleum imports;
   b. Cushion the adverse impact of the high price of energy supplies on consumers, particularly elderly and handicapped low-income persons who cannot afford to make the modifications necessary to reduce their residential energy use; and
   c. Increase, directly and indirectly, job opportunities and national economic output;
4. The primary responsibility for the implementation of such major programs should be lodged with the states. The diversity of conditions among the various states and regions of the nation is sufficiently great that a wholly federally administered program would not be as effective as one that is tailored to meet local requirements and to respond to local opportunities. The state could develop programs using general federal guidelines;
5. To the extent that direct federal administration is more economical and efficient, direct federal financial incentives and assistance should be extended through existing and proven federal programs that would necessitate the creation of separate administrative bureaucracies; and
6. Such new programs should not conflict with existing state energy conservation programs.

Purpose

The basic purpose of this act is to encourage and facilitate the implementation of energy conservation measures and renewable-resource energy measures in dwelling units, nonresidential buildings, and industrial plants through:

1. Supplemental state energy conservation plans; and
2. Federal financial incentives and assistance.

Financial Assistance

Part A of the act deals specifically with weatherization assistance for low-income persons. This assistance was based on congressional findings that:

1. Dwellings owned or occupied by low-income persons frequently are inadequately insulated;
2. Low-income persons, particularly elderly and handicapped low-income persons, can least afford to make the modifications necessary to provide for adequate insulation in such dwellings and to otherwise reduce residential energy use;
3. Weatherization of such dwellings would lower utility expenses for such low-income owners or occupants as well as save thousands of barrels per day of needed fuel oil; and
4. States, through community action agencies established under the Economic Opportunity Act of 1964 and units of general purpose local government, should be encouraged, with the federal financial and technical assistance, to develop and support coordinated weatherization programs designed to ameliorate the adverse effects of high energy costs on such low-income persons to supplement other federal programs serving such persons and to conserve energy.

Implementation

The regulations for weatherization of buildings occupied by low-income persons include provisions:

1. Prescribing, as determined by appropriate federal agencies, for use in various climatic, structural, and human need settings, standards for weatherization materials, energy conservation techniques, and balanced combinations thereof that are designed to achieve a balance of a healthful dwelling environment and maximum practicable energy conservation and designed to assure that:
   a. The benefits of weatherization assistance in connection with leased dwelling units will accrue primarily to low-income tenants;
   b. The rents on such dwelling units will not be raised because of any increased value of the building due to weatherization; and
   c. No undue or excessive enhancement will occur to the value of such dwelling.

Financial Assistance

The financial assistance is to be allocated as to:

1. The number of dwelling units to be weatherized;
2. The climatic conditions of the state; and
3. Other factors that may be important.

Occupational Safety and Health

Occupational Safety and Health Act of 1990 (Williams-Steiger Occupational Safety and Health Act of 1970), Public Law 91-596, December 29, 1970, 84 Statute 1590–1619.

Amended

Public Law 93-237, January 2, 1974, 87 Statute 1024

Public Law 95-251, March 27, 1978, 92 Statute 183

Public Law 97-375, Title I, December 21, 1982, 96 Statute 1821

Public Law 98-620, Title IV, November 8, 1984, 98 Statute 3360

Public Law 101-508, Title III, November 5, 1990, 104 Statute 1388-29

Public Law 102-550, Title X, Subtitle C, October 28, 1992, 106 Statute 3924

Congressional Investigation

This act is based on the findings of Congress that personal injuries and illnesses resulting from working conditions impose a substantial burden upon and are a hindrance to interstate commerce in terms of lost production, wage losses, medical expenses, and disability compensation payments.

Purpose Congress, through its ability to regulate commerce among the states and foreign nations and to secure for all working people safe and healthful working conditions and preserve human resources, declares its purpose and policy:

1. To encourage employers and employees in their efforts to reduce the number of occupational and health

hazards at their places of employment and to stimulate employers and employees to institute new and to perfect existing programs for providing safe and healthful working conditions;

2. To provide that employers and employees have separate but dependent responsibilities and rights with respect to achieving safe and healthful working conditions;
3. To authorize the secretary of labor to set mandatory occupational safety and health standards applicable to businesses affecting interstate commerce and to create an Occupational Safety and Health Review Commission to carry out adjudicatory functions under this act;
4. To build upon advances already made through employer and employee initiatives for providing safe and healthful working conditions;
5. To provide for research in the field of occupational safety and health, including the psychological factors involved and by developing innovative methods, techniques, and approaches for dealing with occupational safety and health problems;
6. To explore ways to discover latent diseases, establishing causal connections between diseases and work in different environmental conditions and conducting other research relating to health problems, in recognition of the fact that occupational health standards present problems often different from those involved in occupational safety;
7. To provide medical criteria that will assure insofar as practicable that no employee will suffer diminished health, functional capacity, or life expectancy as a result of his work experience;
8. To provide for training programs to increase the number and competence of personnel engaged in the field of occupation safety and health;
9. To provide for the development and promulgation of occupational safety and health standards;
10. To provide an effective enforcement program that shall include a prohibition against giving advance notice to any inspection and sanctions for any individual violating this prohibition;

11. To encourage the states to assume the fullest responsibility for the administration and enforcement of their occupational safety and health laws;
12. To provide for appropriate reporting procedures with respect to occupational safety and health; and
13. To encourage joint labor management efforts to reduce injuries and disease arising out of employment.

Standards

A National Institute for Occupational Safety and Health is to be established to prepare standards:

1. Standards are to deal with toxic materials or harmful physical agents that most adequately assure to the extent possible, on the basis of the best available evidence, that no employee will suffer impairment of health or functional capacity; and
2. The standards are to be based upon research, demonstrations, experiments, and such other information as may be appropriate. In addition to the attainment of the highest degree of health and safety protection for the employee, other considerations shall be the latest available scientific data in the field, the feasibility of the standards, and experience gained under this and other health and safety laws.

Implementation

To implement the act, a National Advisory Committee on Occupational Safety and Health was established. The committee advises, consults with, and makes recommendations to the secretary of labor and the secretary of health, education, and welfare.

An inspection of businesses and industries is to collect evidence of infractions of the act. A penalty of up to $10,000 may be imposed for infractions.

Research Program

The act also provides that the secretary of labor and secretary of health, education, and welfare shall conduct research, experiments, and demonstrations relating to occupational safety and health. As part of the secretaries' activities, a list of all known toxic substances by generic family is to be collected along with the concentrations at which points toxicity is known to occur. This report will be published annually.

Toxic Substances

Toxic Substances Control Act of 1976, Public Law 94-469, October 17, 1976, 90 Statute 2003.

Amended

Public Law 94-465, October 11, 1976, 90 Statute 2003

Public Law 97-129, December 29, 1981, 95 Statute 1686

Public Law 98-80, August 23, 1985, 97 Statute 485

Public Law 98-620, Title IV, November 8, 1984, 98 Statute 3358

Public Law 99-419, October 22, 1986, 100 Statute 954

Public Law 100-368, July 18, 1988, 102 Statute 829

Public Law 100-418, Title I, August 23, 1988, 102 Statute 829

Public Law 100-551, October 28, 1988, 102 Statute 2755

Public Law 101-508, Title X, November 5, 1990, 104 Statute 4593; 4596, 4597

Public Law 102-550, Title X, October 28, 1992, 106 Statute 3912 to 3923

Public Law 103-382, Title III, Part I. October 20, 1994, 108 Statute 4022, 4026

Purpose

The basic purpose of this act is to prevent unreasonable risks of injury to health or the environment during the manufacture, processing, distribution, use, or disposal of chemical substances. The chemical industry has grown tremendously in the last century. It is estimated that currently there are 2 million recognized chemical compounds, and there are about 250,000 new compounds produced each year. The Environmental Protection Agency estimates that about 1,000 of these new chemicals will find their way into the marketplace and subsequently into the environment through use or disposal.

We use chemicals in a majority of our daily activities. As a result of wearing, washing, inhaling, and ingesting we come into contact with a multitude of chemical substances. Most of these chemicals do not affect human health but often present health and environmental dangers.

Regulations

In 1991 the Council on Environmental Quality concluded that regulatory measures to control toxic chemicals were inadequate. This report provided the impetus for the Toxic Substance Control Act. This legislation has evolved into a comprehensive program to protect the public and the environment from exposure to hazardous chemicals.

Russell E. Train, former administrator of the Environmental Protection Agency, has stated:

> Most Americans had no idea, until relatively recently, that they were living so dangerously. They had no idea that when they went to work in the morning, or when they ate their breakfast—that when they did the things they had to do to earn a living and keep themselves alive and well—that when they did things as ordinary, as innocent and essential to life as eating, drinking, breathing or touching, they could, in fact, be laying their lives on the line. They had no idea that, without their knowledge or consent, they were often engaging in a grim game of chemical roulette whose results they would not know until many years later.

Protection of Health and Environment

This act provides the means to protect health and the environment before a new chemical is manufactured for commercial use. This is the most effective and efficient time to prevent unreasonable risks to public health and the environment. It is at this point that the costly regulation in terms of human suffering, jobs lost, wasted capital expenditures, and other costs are lowest. To illustrate, vinyl chloride, a chemical used to produce plastics, has been implicated in causing liver cancer in industrial workers. The use of plastics has grown rapidly in recent years. Vinyl chloride is now regulated, but the cost was high. If the dangers had been known earlier, alternatives could have been developed, and polyvinyl chloride plastics might not have become so important in the economy of the nation before being banned.

Clean Air

Clean Air Act of 1955, Chapter 360, July 14, 1955, 69 Statute 322. Also known as the Air Pollution Control Act.

Major Amendments

Clean Air Act Amendments of 1966, Public Law 89-675, October 15, 1966, 80 Statute 954

Air Quality Act of 1967, Public Law 90-148, November 21, 1967, 81 Statute 485

Clean Air Amendments of 1970, Public Law 91-604, December 31, 1970, 84 Statute 1676 to 1713

Clean Air Act Amendments of 1977, Public Law 95-95, August 7, 1977, 91 Statute 685

Public Law 95-190, November 16, 1977, 91 Statute 1404, 1405

Public Law 97-375, Title I, December 21, 1982, 96 Statute 1820

Clean Air Act Amendments of 1990, Public Law 101-549, November 15, 1990, 104 Statute 2399

Amendments

Public Law 86-365, September 1959, 73 Statute 646

Public Law 88-206, December 17, 1963, 77 Statute 392

Public Law 89-272, Title I, October 20, 1965, 79 Statute 992

Public Law 91-137, December 5, 1969, 83 Statute 283

Public Law 92-157, Title III, November 18, 1971, 85 Statute 464

Public Law 93-15, April 9, 1973, 85 Statute 464

Public Law 93-319, June 22, 1974, 86 Statute 248 to 259, 261, 265

Public Law 96-300, July 2, 1980, 94 Statute 831

Public Law 97-23, July 17, 1981, 95 Statute 139

Public Law 98-213, December 8, 1983, 97 Statute 1461

Public Law 102-187, December 4, 1991, 105 Statute 1285

Public Law 103-437, November 2, 1994, 108 Statute 4594

Federal clean air legislation began with the Clean Air Act of 1955. As the problem was redefined, the Clean Air Act has been amended many times, the last time in 1994. The 1955 act provided the basic policy for the nation. The original act states:

> That in recognition of the dangers to the public health and welfare, injury to agricultural crops and livestock, damage to and deterioration of property, and hazards to air and ground transportation, from air pollution, it is hereby declared to be the policy of Congress to preserve and protect the primary responsibilities and rights of the states and local governments in controlling air pollution, to support and aid technical research, to devise and develop methods of abating such pollution, and to provide federal technical services and financial aid to state and local agencies and institutions in the formulation and execution of their air pollution abatement research programs.

The surgeon general of the Public Health Service under the direction of the secretary of health, education, and welfare, and with help from other governmental agencies, is to implement the act. It includes:

1. Encouragement of cooperative activities of state and local governments for the prevention and abatement of air pollution;
2. Collection and dissemination of information on air pollution;
3. Devising and developing methods of preventing and abating air pollution; and
4. Making available to state and local air pollution agencies results of surveys, studies, investigations, research and experiments relating to air pollution.

Noise

Noise Control Act of 1972, Public Law 92-574, October 27, 1972, 86 Statute 1234.

Congressional Investigation

In a study of the problems of noise, Congress found that:

1. Inadequately controlled noise presents a growing danger to the health and welfare of the nation's population, particularly in urban areas;
2. The major sources of noise include transportation vehicles and equipment, machinery, appliances, and other products in commerce;
3. Although primary responsibility for control of noise rests with state and local governments, federal action is essential to deal with the major noise sources in commerce, control of which requires national uniformity of treatment.

Purpose

Because it is the responsibility of Congress to develop policies to promote an environment for all Americans free from noise that jeopardizes their health or welfare, it is the purpose of this act to establish a means for effective coordination of federal research and activities in noise control to authorize the establishment of federal noise emission standards for products distributed in commerce and to provide information to the public respecting the noise emission and noise reduction characteristics of such products.

Identification of Major Noise Sources

This act specifies that the Environmental Protection Agency was to develop and publish criteria with respect to noise. Such criteria were to reflect the scientific knowledge most useful in indicating the kind and extent of all identifiable effects on the public health or welfare that may be expected from differing quantities and qualities of noise. This information is to be published, identifying products that are major sources of noise and also giving information on techniques for control of noise from such products, including available data on the technology, costs, and alternative methods of noise control.

Noise Emission Standards

The act requires that the Environmental Protection Agency establish noise emission standards. These noise emission standards are to apply to the following categories:

1. Construction equipment;
2. Transportation equipment (including recreational vehicles and related equipment);
3. Any motor or engine (including any equipment of which an engine or motor is an integral part); and
4. Electrical or electronic equipment.

Labeling

In order to control noise the EPA has the responsibility to label products that:

1. Emit noise capable of adversely affecting the public health or welfare; or
2. Are sold wholly or in part on the basis of effectiveness in reducing noise.

The act addresses specifically the problems of noise from:

1. Aircraft
2. Railroads
3. Motor carriers

Low-Noise Emission Products

The EPA has been given the authority to establish a Low-Noise Emission Product Advisory Committee to assist in determining which products qualify as low-noise emission products.

Quiet Communities Act of 1978, Public Law 95-609, November 8, 1978, 92 Statute 3079.

Purpose

This act amends the Noise Control Act of 1972. It has the same objectives of the original act: to develop effective federal noise control programs. Its basic purpose is to:

1. Develop and disseminate information and educational materials to all segments of the public on the public health and other effects of noise and the most effective means for noise control, through the use of materials for school curriculums, volunteer organizations, radio and television programs, publications, and other means;
2. Conduct research directly or with any public or private organization or any person on the effects, measurement, and control of noise, including:
   a. Investigation of the psychological and physiological effects of noise on humans, domestic animals, wildlife, and property, and the determination of dose/response relationships suitable for use in decision making, with special emphasis on the nonauditory effects of noise;
   b. Investigation, development, and demonstration of noise control technology for products subject to possible regulation;
3. The investigation, development, and demonstration of monitoring equipment and other technology especially suited for use by state and local noise control programs;
4. The investigation of the economic impact of noise on property and human activities; and
5. The investigation and demonstration of the use of economic incentives (including emission charged in the control of noise).

Quiet Communities Program

A program was to be established to:

1. Give grants to states, local governments, and authorized regional planning agencies for the purpose of:
   a. Identifying and determining the nature and extent of noise problems;
   b. Planning, developing, and establishing a noise control capacity;
   c. Developing abatement plans; and
   d. Evaluating techniques for controlling noise and demonstrating the best available techniques;

2. Purchase monitoring and other equipment;
3. Develop and implement a quality assurance program for equipment and monitoring procedures;
4. Conduct studies and demonstrations to determine the resources and personal needs of the states and local governments;
5. Develop educational and training materials and programs including national and regional workshops to support noise abatement and control programs;
6. Establish regional technical assistance centers; and
7. Provide technical assistance.

Directory of Organizations 3

Organizations that consider the problem of indoor pollution can be listed in three categories. The first category includes the principal agencies of the U.S. government. These agencies play an important role in developing the basic pollution policies of the nation. The second group includes a number of intergovernmental advisory committees that have specific, limited objectives. The third group, national organizations, provides information on specific types of pollutants. This chapter concludes with a list of organizational sources, which are revised annually.

Government Organizations of the United States

Environmental Protection Agency (EPA)
401 M Street SW
Washington, DC 20460
Phone: 202-260-2090

Description: EPA was established in the executive branch as an independent agency on December 2, 1970. It was created to provide coordinated and effective governmental action on behalf of the environment. The agency is designed to serve as the public's advocate for a livable environment.

Purpose: Its major activity is to administer programs in the areas of air and radiation, water, solid wastes, and the control of pesticides and toxic substances.

Indoor Pollution Programs: The EPA has no specific programs to control indoor pollution. Indirectly and frequently, however, EPA influences the quality of the indoor atmosphere. For example, outdoor air quality affects the quality of the air indoors, and it is frequently impossible to distinguish the effects of toxic substances indoors or outdoors. Specific EPA air monitoring programs are:

Interagency Committee on Indoor Air Quality (CIAQ)
Indoor Air Division Staff
Environmental Protection Agency
401 M Street SW
Washington, DC 20460
Phone: 202-260-2090

Description: CIAQ was established under authority of the Superfund Amendments and Reauthorization Act of 1986 (SARA). It is an interagency committee composed of representatives of federal agencies and departments concerned with various aspects of indoor air quality.

Purpose: CIAQ deals with all types of indoor air pollution and coordinates federal research and policy on indoor air activities, including radon, allergens in indoor air, environmental tobacco smoke, sick building syndrome, indoor carcinogens, biocontaminate control, toxic chemicals/household products, chemicals emitted from carpets, formaldehyde and pressed wood products, pesticides, and drinking water.

Publications and Reports: Current Federal Indoor Air Quality Activities. Available from EPA, Public Information Center, 401 M Street SW, Washington, DC 20460

Meetings: Quarterly.

National Air Pollution Control Techniques Advisory Committee
Emission Standards Division
Office of Air Quality Planning and Standards
Environmental Protection Agency
Research Triangle Park, NC 27711

Description: Established by the surgeon general on March 4, 1968, under the authority of Section 110(d) of the Clean Air Act. Reestablished by the administrator of the Consumer Protection and Environmental Health Service, pursuant to the Secretary's Reorganization Order, July 1, 1968; transferred to EPA pursuant to Reorganization Plan No. 3, December 2, 1970; and reconstituted by the administrator of EPA, June 8, 1971, pursuant to Sections 108(b)(1) and 2 and 117(f) of the Clean Air Act Amendments of 1970. Members consist of director, Emission Standards Division, who serves as chair. Eleven members are appointed by the EPA administrator of EPA for one- to three-year terms. Additional members are selected from the chemical, engineering, biomedical, legal, environmental, and socioeconomic disciplines in universities, state and local governments, research institutions, and industry. Members are selected for their technical expertise and interest in developing air pollution control techniques. Meetings are called as needed by the chair.

Purpose: To control contamination of air quality.

Activities: Acts as advisory committee of the Emission Standards Division, Office of Air Quality Planning and Standards, Office of Air and Radiation, Environmental Protection Agency. Advises on latest available technology and methods to control air contamination to be published in air quality control documents. Also monitors testing methods for categories of new sources of air pollutants.

Publications: Documents on air quality techniques.

Radon Workgroup
Environmental Protection Agency
401 M Street SW
Washington, DC 20460
Phone: 202-260-2090

Description: The Radon Workgroup, established in 1986, is an interagency group of the Committee on Indoor Air Quality. The group consists of representatives from the Department of Energy and EPA.

Purpose: Provides information to CIAQ on pending federal radon legislation, ongoing programs, and other issues. It coordinates activities related to radon and develops federal responses to radon-related issues, including identifying and discussing technical and nontechnical interagency issues, reviewing member agency research and program plans, planning and implementing joint projects and coordinating multiagency participation in radon-related activities, and producing and publishing information related to radon activities.

U.S. Department of Energy (DOE)
1000 Independence Avenue SW
Washington, DC 20585
Phone: 202-586-5000

Description: The Department of Energy was established by the Department of Energy Organization Act on October 1, 1977. The act consolidated the major federal energy functions into a single Cabinet-level department.

Purpose: DOE provides the framework for a comprehensive and balanced national energy plan through the coordination and administration of the energy functions of the federal government.

Indoor Pollution Programs: DOE does not have any programs that deal directly with indoor pollution. Indirectly, through energy conservation programs, it has affected the quality of air in houses and other buildings. Through building codes, insulation has been increased, resulting in the use of less energy to heat buildings. At the same time, however, the exchange of air in buildings is reduced, increasing toxic impurities. This has created a health problem in buildings known as sick building syndrome.

U.S. Department of Health and Human Services (HHS)
200 Independence Avenue SW
Washington, DC 20201
Phone: 202-689-0257

Description: The Department of Health, Education, and Welfare was created on April 11, 1953. This department was redesignated as the Department of Health and Human Services on May 4, 1980, by the Department of Education Organization Act of 1980.

Purpose: HHS's basic purpose is to assure that the health and human services of the nation are protected.

Indoor Pollution Programs: There is no specific agency within the department that considers the problems of indoor pollution. A number of the agencies in their broad objectives touch on the problems of the effects of indoor pollution on health.

The Centers for Disease Control and Prevention administers national programs for the prevention and control of communicable and vector-borne diseases. It develops and implements programs to address environmental health problems, including responses to environmental, chemical, and radiation emergencies. The agency's National Health Promotion Program includes research, information, and education in the field of smoking and health.

As part of its responsibility, the Agency for Toxic Substances and Disease Registry develops programs to protect the public and workers from exposure to and adverse health effects of hazardous substances released into the environment. It assists EPA in identifying hazardous waste substances to be regulated. It also develops scientific and technical procedures for evaluating public health risks from hazardous-substance incidents and for developing recommendations to protect public and workers when exposed to hazardous substances.

The National Center for Toxicological Research conducts research programs to study the biological effects of potentially toxic chemical substances found in the environment, emphasizing the determination of the health effects resulting from long-term, low-level exposure to chemical toxicants and the basic biological processes for chemical toxicants in animal organisms.

U.S. Department of Housing and Urban Development (HUD)
451 Seventh Street SW
Washington, DC 20410
Phone: 202-708-1422

Description: HUD was established by the Department of Housing and Urban Development Act, November 9, 1965.

Purpose: The department was created to:

1. Administer the principal programs that provide assistance for housing and for development of the nation's communities;

2. Encourage the solution of housing and community development problems through states and localities; and
3. Encourage the maximum contributions that may be made by various private homebuilding and mortgage lending industries, both primary and secondary, to housing, community development, and the national economy.

Indoor Pollution Programs: The Office of Lead-Based Paint Abatement and Poisoning Prevention is responsible for all lead-based paint abatement and poisoning prevention activities within HUD, including policy development, abatement, training, regulations, and research. Specific activities include:

1. Providing awareness to the public and building industry of the dangers of lead-based paint poisoning and the options for detection, risk reduction, and abatement;
2. Encouraging the development of safer, more effective, and less costly methods for detection, risk reduction, and abatement; and
3. Encouraging state and local government to develop lead-based paint programs covering primary prevention, including public, education, contractor certification, hazard reduction, financing, and enforcement.

U.S. Department of Labor (DOL)
200 Constitution Avenue NW
Washington, DC 20210
Phone: 202-259-5000

Description: The Bureau of Labor was first created by Congress in 1884 under the Interior Department. The bureau later became the independent Department of Labor but did not enjoy executive rank. On February 14, 1903, it again returned to bureau status within the Department of Commerce and Labor. The current Department of Labor was created by congressional act on March 4, 1913.

Purpose: The purpose of the Department of Labor is to foster, promote, and develop the welfare of wage earners to improve working conditions and to advance their opportunities for profitable employment.

Indoor Pollution Programs: DOL administers a variety of federal labor laws guaranteeing workers' rights to safe and healthy working conditions; other considerations are minimum hourly wage and overtime pay, freedom from employment discrimination, employment insurance and workers' compensation, and other items.

The Occupational Safety and Health Administration (OSHA) within the Department of Labor was established in 1970 to develop occupational safety and health standards and to conduct investigations and inspections to determine the status of compliance with safety and health standards and regulations and to issue citations and propose penalties for noncompliance with standards and regulations.

The Office of the American Workplace was created on July 27, 1993, and is responsible for administering and directing workplace programs that encourage the development of work organizations, human resource practices, technology, and performance measurements that enhance business competitiveness.

Associations

Adhesive and Sealant Council (ASC)
1627 K Street NW, Suite 1000
Washington, DC 20006-1707
Phone: 202-452-1500
Fax: 202-452-1501

Description: Founded 1957. Has 184 members and nine staffpeople. Members are manufacturers and people who sell adhesives and sealants and raw material suppliers. Budget: $1.3 million. Holds annual conference in June at Hilton Head and semiannual conventions in March/April and October.

Purpose: To promote education on adhesives and sealants.

Activities: Has consulting firms. Operates Adhesive and Sealant Council Education Foundation. Compiles statistics. Has library on technical subjects. Presents the ASC Award annually to recognize a person who contributes most to industry technology and civic programs. Has several committees.

Publications: ASC Journal, periodic; *Catalyst,* monthly; *Guide to Hot Melt Systems; Sealant Specifications; Memberships Directory,* annual.

Air and Waste Management Association (AWMA)
1 Gateway Center, 3rd Floor
Pittsburgh, PA 15222
Phone: 412-232-3444
Fax: 412-232-3450

Description: Founded 1907. Has 15,000 members and 45 staffpeople and is multinational. Membership dues are $85 annually, $25 (student), $42 (affiliate), and $1,000 (corporation). Budget is $9,458,000. There are 25 sections and 56 chapters. Members are environmentalists. Holds annual meeting in June. Has periodic symposia, conferences, and workshops. The Air Pollution Control Association is a division of the Air and Waste Management Association.

Purpose: Disseminate information on environmental problems.

Activities: Acts as an environmental, educational, and technical organization. Exchanges technical information on environmental topics. Has a reference library containing materials on air pollution, environmental management, and the like. Presents several awards annually—Frank A. Chambers, Lyman A. Ripperton, J. Deane Gensenbaugh, Richard Beatty Mellon, S. Smith Griswold—for special achievements. Has several committees.

Publications: Environmental Manager, monthly; *Journal of the Air and Waste Management Association,* monthly; *Manuals; Proceedings; Proceedings Digest,* annual; *Resource Book and Membership Directory,* annual. Also publishes summary guides and educational materials.

Air Movement and Control Association (AMCA)
30 West University Drive
Arlington Heights, IL 60004-1893
Phone: 708-394-0150
Fax: 708-253-0088

Description: Founded 1955. Has 218 members and 23 staffpeople. Members are manufacturers and people interested in air moving and control equipment. Budget is $2 million. Holds annual meeting and periodic meetings.

Purpose: To promote air movement and control equipment.

Activities: Conducts research on improving methods of testing equipment. Develops codes for fans, louvers, dampers, and so on. Operates testing laboratory. Formerly the Air Moving and Conditioning Association.

Publications: AMCA Newsletter, quarterly; *Directory of Licensed Products,* annual; *Engineering Newsletter,* quarterly; *Manuals, Member Handbook,* annual; *Techspecs,* semiannual. Also publishes standards and application guides.

Alliance to End Childhood Lead Poisoning (AECLP)
227 Massachusetts Avenue NE, Suite 200
Washington, DC 20002

Description: Founded 1990. Has 11 staffpeople. Nonmembership. Budget is $1.2 million. Holds conferences.

Purpose: To end childhood lead poisoning.

Activities: Seeks to inform the public, health officials, and policymakers of the hazards posed by lead and the need to prevent lead poisoning. Works to strengthen federal programs and policies developing cost-effective strategies for protecting children and developing a national infrastructure for reducing lead hazards. Keeps pressure on federal agencies, overcomes private-sector obstacles, and mobilizes other resources. Conducts lobbying activities. Has a Technical Advisory Committee.

Publications: Alliance Alert, bimonthly. Also publishes brochures and technical, policy, and program guidelines.

American Association of Textile Chemists and Colorists (AATCC)
P.O. Box 12215
Research Triangle Park, NC 27709-2215
Phone: 919-549-8141
Fax: 919-549-8933

Description: Founded 1921. Has 7,956 members and 24 staffpeople. Membership dues: individual $55 (annual). There are five regional granges and 13 local groups. Members are textile chemists and colorists. Is multinational. Holds annual meeting.

Purpose: To disseminate information to textile chemists and colorists.

Activities: Test methods such as colorfastness to light, washing, perspiration, crease resistance, durable press, water resistance, and so on. Develops standard test methods. Conducts textile test method research. Has 66 research committees. Has speakers' bureau. Has library, open to public. Presents the Olney Award annually.

Publications: AATCC Bank of Papers, annual; *AATCC Membership Directory,* annual; *AATCC Technical Manual,* annual; *Buyer's Guide,* annual; *Handbook, Symposium Papers,* periodic; *Textile Chemist and Colorist,* monthly.

American Conference of Governmental Industrial Hygienists (ACGIH)
1330 Kemper Meadow Road
Cincinnati, OH 45240
Phone: 513-742-2020
Fax: 513-742-3355

Description: Founded 1938. Has 5,500 members and 23 staffpeople. Membership dues $65. Budget is $3.1 million. Is multinational. Members are professionals employed by the government responsible for programs of industrial hygiene, educators, and those conducting research in industrial hygiene. Holds conference and exposition convention (exhibits), always May in Washington, DC.

Purpose: To protect industrial workers' health and safety.

Activities: Acts as a medium for exchange of ideas and promotes standards and techniques in industrial health. Compiles statistics. Conducts educational programs. Library has 2,000 books, periodicals, and archival material on industrial hygiene and occupational and environmental health and safety. Bestows the Herbert E. Stockinger Award annually contributions to the field of industrial and environmental toxicology.

Publications: Applied Occupational and Environmental Hygiene, monthly; *Industrial Ventilation—A Manual of Recommended Prac-*

tice; Guidelines for the Assessment of Bioaerosols in the Indoor Environment; Air Sampling Instruments; Threshold Limit Values and Biological Exposure Indices; Ventilation System Testing. Also publishes other manuals, guides, and studies.

American Consulting Engineers Council (ACEC)
1015 15th Street, NW, Suite 803
Washington, DC 20005
Phone: 202-347-7474
Fax: 202-898-0068

Description: Founded 1973. Has 5,500 members, who include consulting engineering firms. Budget is $4.5 million. There are 51 regional groups. Holds semiannual convention.

Purpose: To promote private consulting engineering programs.

Activities: Conducts programs concerned with business practices, government affairs, public relations, and professional liability. Compiles statistics on office practices, employment, insurance, and services provided. Holds seminars and educational programs. Maintains speakers' bureau. Presents Engineering Excellence Award annually.

Publications: American Consulting Engineer, quarterly; *Interpro,* monthly; *The Last Word,* weekly; *Minority Directory,* periodic; *Membership Directory,* annual.

American Industrial Hygiene Association (AIHA)
475 Wolf Ledges Parkway
Akron, OH 44311-1081

Description: AIHA was founded in 1939.

Purpose: To promote the study and control of environmental factors affecting the health and well-being of industrial workers.

Activities: Although the major goal is to produce a healthy industrial environment, many concepts are transferable to commercial and residential environments. The association has a laboratory accreditation program based on participation in the National Institute for Occupational Safety and Health's proficiency testing program. This includes bimonthly samples for lead, cadmium, zinc, asbestos, silica, and organic solvents. Among AIHA's technical committees is one on indoor environmental quality.

Publication: Publishes a monthly *American Industrial Hygiene Association Journal* and maintains a list of occupational laboratories.

American Institute of Architects (AIA)
1735 New York Avenue NW
Washington, DC 20006
Phone: 202-626-7300
Fax: 202-626-7421

Description: Founded 1857. Has 57,000 members and 225 staffpeople. Budget is $32 million. There are 19 regional groups, 50 state groups, and 301 local groups. Members are professional architects. Holds annual meeting.

Purpose: To promote architecture as a profession.

Activities: Has education and training programs. Promotes design excellence by influencing change. Sponsors educational programs with schools of architecture. Conducts development programs in architecture. Acts as adviser on professional competition. Established the American Architectural Foundation. Sponsors Octagon Museum. Holds exhibitions. Compiles statistics. Has speakers' bureau. Conducts research programs, charitable activities, and children's services. Has library of 3,000 books, artwork, and audiovisual materials. Presents Gold Medal Award for achievement in design and the National Honor Award.

Publications: American Institute of Architects Memo, monthly; *ARCHITECTURE,* monthly; *Profile,* annual; *Membership Directory.*

American National Standards Institute (ANSI)
11 W. 42nd Street, 13th Floor
New York, NY 10036
Phone: 212-642-4900
Fax: 212-398-0023

Description: Founded 1918. Has 1,600 members and 107 staffpeople. Budget is $16 million. Membership includes industrial firms, trade associations, technical societies, labor organizations, government agencies, and consumer organizations. Holds an annual conference.

Purpose: To provide health and safety conditions in building construction for those using the buildings.

Activities: Serves as a clearinghouse for nationally coordinated standards for building construction. Confers American National Standards status to standards developed in such areas as materials, methods of testing and analysis, safety, health, and building construction. Provides information on foreign standards and represents the U.S. interest in international standards. Presents ANSI Award annually. Does much work through committees and councils.

Publications: ANSI Reports, monthly; *Catalog of American National Standards,* annual; *Standards Action,* biweekly.

American Public Health Association (APHA)
1015 15th Street, NW
Washington, DC 20005
Phone: 202-789-5600
Fax: 202-789-5681

Description: Founded 1872. Has 31,500 members and 65 staffpeople. Has $9.6 million budget. Members are physicians, nurses, social workers, and anyone interested in health. Holds annual meetings.

Purpose: To promote information to improve public health.

Activities: Protects and promotes personal, mental, and environmental health. Establishes uniform standards and procedures. Conducts research in public health. Explores medical care programs and their relationships to public health. Sponsors job placement service. Has library (not open to public). Presents awards for excellence, such as the Drotman Award and the Sedgwick Memorial Award.

Publications: American Journal of Public Health, monthly; *The Nation's Health,* ten issues per year; also publishes books, manuals, and pamphlets.

American Society for Testing and Materials (ASTM)
1916 Race Street
Philadelphia, PA 19103-1187
Phone: 215-299-5400
Fax: 215-977-9679

Description: Founded 1898. Has 35,000 members and 200 staffpeople. Membership dues are $65 for an individual and $350 for an organization. Budget is $24 million. Is multinational. Members are scientists, managers, academicians, skilled technicians, government agencies, and laboratories. Holds monthly meetings. Holds 40 symposia each year.

Purpose: To test and promote materials for buildings to provide proper ventilation and air quality.

Activities: Establishes standards for materials, products, systems, and services. Has 131 technical committees. New committees are formed periodically to keep up-to-date with new developments. Has developed more than 9,000 standard test methods, specifications, and recommended practices now in use.

Publications: Annual Book of ASTM Standards; Cement, Concrete, and Aggregates Journal, semiannual; *Geotechnical Testing Journal,* quarterly; *Journal of Testing and Evaluation,* bimonthly; *Standardization News,* monthly. Also publishes papers and reports.

American Society of Heating, Refrigeration, and Air-Conditioning Engineers (ASHRAE)
1791 Tullie Circle NE
Atlanta, GA 30329
Phone: 404-636-8400
Fax: 404-321-5478

Description: Founded 1894. Has 50,000 members and 95 staffpeople. Annual dues are $115. Budget is $12 million. There are 12 regional groups and 157 local groups. Holds annual winter meeting and exposition.

Purpose: To study effects of air-conditioning, heating, and ventilating on indoor air quality.

Activities: Sponsors research programs in cooperation with universities, research laboratories, and government agencies on indoor air quality, heat transfer, flow, and cooling processes. Conducts research and general technical programs through 90 technical committees. Conducts professional development seminars. Has a reference library. Presents Telecommunication Services Award. Operates through several committees, councils, and special sections.

Publications: ASHRAE Insights, monthly; *ASHRAE Journal,* monthly; *ASHRAE Transactions,* semiannual; books; four-volume handbook; proceedings, reports, and engineering standards.

Asbestos Information Association/
North America (AIA/NA)
1745 Jefferson Davis Highway, Suite 406
Arlington, VA 22202
Phone: 703-412-1150
Fax: 703-412-1152

Description: Founded 1970. Has 17 members and two staffpeople. Budget is $260,000. Members are manufacturers, processors, miners, and millers of asbestos or products containing asbestos. Holds annual conference.

Purpose: To provide information on how asbestos affects health.

Activities: Works with government agencies to develop and implement standards regarding exposure to asbestos dust and to control emissions of asbestos dust. Helps solve problems arising from the health effects of asbestos. Acts as a central agency for the collection and dissemination of medical and technical information on asbestos-related disease and asbestos dust control. Has reference library.

Publications: News and Notes, monthly. Also publishes information/technical materials.

Association of the Walls and Ceiling
Industries International (AWCII)
307 E. Annandale Road, Suite 200
Falls Church, VA 22042-2433
Phone: 703-534-8300
Fax: 703-534-8307

Description: Founded 1976. Has 950 members and 12 staffpeople. Members include those working with acoustical tile, plastering, asbestos abatement, and insulation. Budget is $1.7 million. Is multinational. Holds annual meeting.

Purpose: To keep members informed about changes in the profession.

Activities: Keeps informed about improvements and changes that occur about acoustical tile, asbestos abatement, fireproofing, insulation, and the like. Presents a recognition award. Has a library of 3,000 holdings of books, periodicals, and videos. Has several committees.

Publications: Buyers Guide for the Wall and Ceiling Industry, annual; *Construction Dimensions Magazine,* monthly; *Information Resources Catalog; Technical Information; Who's Who in the Wall and Ceiling Industry,* annual.

Association of University Programs in Occupational Health and Safety
1600 Clifton Road
Atlanta, GA 30333
Phone: 404-639-3771

Description: Founded 1977. Has 14 members. Universities offering graduate training and continuing education for occupational health and safety professionals. Holds semiannual meetings.

Purpose and Activities: Provides a forum for the exchange of information among members on graduate training in occupational medicine, industrial hygiene, and industrial safety engineering. Works in conjunction with the National Institute for Occupational Safety and Health to facilitate the operation of training programs.

Publications: Newsletter, periodic.

Building Officials and Code Administrators International (BOCA)
4051 West Flossmoor Road
Country Club Hills, IL 60478-5795
Phone: 708-799-2300
Fax: 708-799-4981

Description: Founded 1915. Has 14,500 members and 65 staffpeople. Budget is $8 million. Members are those interested in administering or formulating building codes and housing regulations. Holds midwinter meeting.

Purpose: To promote and keep up-to-date on building codes.

Activities: Promotes establishment of unbiased building codes and keep them up-to-date. Gives information on quality of building materials and new construction methods. Maintains service for local governments on building codes and enforcement. Does consulting. Holds seminars on training and education regarding building codes. Has correspondence courses. Maintains placement services for several committees. Presents the Albert H. Baum and the Walter S. Lee Awards annually.

Publications: Annual Supplements to National Codes; BOCA Bulletin, bimonthly; *Building Code Manual and Property Management Manual; Building Official and Code Administrator Magazine,* bimonthly; *Membership Directory,* annual; *National Building Code,* triennial; *National Energy Conservation Code,* triennial; *National Fire Prevention Code,* triennial; *National Mechanical Code,* triennial; *National Plumbing Code,* triennial; *National Property Management Code,* periodic; *One-and-Two Family Dwelling Code; Plumbing Manual; Proposed Code Changes,* annual; *Research Report,* quarterly. Also publishes code interpretations, permits, department forms, and textbooks.

Building Owner and Manager
Association International (BOMA)
1201 New York Avenue NW, Suite 300
Washington, DC 20005
Phone: 202-408-2662
Fax: 202-371-0181

Description: Founded 1908. Has 15,500 members and 40 staffpeople. Members are owners, managers, investors, and developers of commercial office buildings. Budget is $4.8 million. Has ten regional groups, six state groups, and 101 local groups. Holds annual convention and trade show.

Purpose: To keep managers, owners, and developers of commercial office buildings informed of recent developments.

Activities: Promotes the office building industry through discussion, education, and cooperation. Provides courses leading to RPA (real property administrator), SMA (systems maintenance administrator), and FMA (facilities maintenance administrator) certifications. Conducts research programs. Operates placement service. Maintains speakers' bureau. Compiles statistics. Maintains 20 sections, such as Accounting Systems and Procedures, Legislative and Urban Affairs, and Corporate and Financial Buildings. Has a reference library. Presents Office Building of the Year Award annually.

Publications: Buyer's Guide, annual; *Cleaning Study, BOMA Office Market Review,* annual; *Experience Exchange Report for Downtown and Suburban Office Buildings,* annual; *Functional Accounting Guide and Chart of Accountants, Membership Directory,* annual; *Office Tenant Moves and Changes; Skylines,* monthly. Also publishes educational texts, workbooks, and analyses.

Carpet and Rug Institute (CRI)
P.O. Box 2048
Dalton, GA 30722
Phone: 706-278-3176
Fax: 706-278-8835

Description: Founded 1968. Has 280 members and 24 staffpeople. Members belong to the trade association for manufacturers of carpets, rugs, bath mats, and the like. Holds annual meetings.

Purpose: To promote relationship among manufacturers of rugs, carpets, bath mats, and bedspreads.

Activities: Serves as trade association for manufacturers of carpets. Associate members provide raw materials and services to the industry. Conducts programs in the areas of standards, market and technical services, and public relations. Compiles statistics on shipments, dollar value, and materials used. Has several committees.

Publications: Carpet and Rug Industry Review, annual; *Carpet and Rug Manufacturing Plants and Corporate Locations in the U.S. and Canada,* annual; *Carpet Specifiers Handbook; A Complete Commercial Carpet Reference,* manual; *CRI Membership Directory,* annual.

Chemical Industry Institute of Toxicology (CIIT)
P.O. Box 12137
Research Triangle Park, NC 27709
Phone: 919-558-1200
Fax: 919-558-1300

Description: Founded 1974. Has 47 members and 160 staffpeople. Members are chemical and pharmaceutical companies.

Purpose: To develop scientific data needed to evaluate potential risks of chemicals and pharmaceutical products.

Activities: Studies health risks from occupational exposure. Improves products used in safety evaluation. Updates toxicological testing. Has advisory panel. Provides assistantships in graduate and postdoctoral toxicological training. Conducts workshops. Has reference library of 25,000 holdings. Bestows Founders Award annually. Has many departments.

Publications: CIIT Activities, monthly.

Construction Specifications Institute (CSI)
501 Madison Street
Alexandria, VA 22314-1791
Phone: 703-684-0300
Fax: 703-684-0465

Description: Founded 1948. Has 17,000 members. Membership dues $130 annually. Budget is $8.5 million. There are ten regional groups and 128 local groups. Members are architects, professors, contractors, and anyone interested in specifications and documents used for construction projects. Holds annual convention.

Purpose: To keep those interested informed about any specifications used in their field.

Activities: Advances construction technology through education, research, and communication. Certifies construction specifiers involved in construction. Maintains speakers' bureau. Offers seminars. Has reference library. Presents recognition awards.

Publications: The Construction Specifier, monthly; *Membership Directory,* annual; *The Newsdigest,* monthly. Also publishes technical documents and monographs.

Consumer Product Safety Commission (CPSC)
5401 Westbard Avenue
Washington, DC 20207-0001
Phone: 301-504-0580

Description: Founded 1973. Has a staff of 32 professional scientists.

Purpose and Activities: Conducts studies to determine exposure to hazardous chemicals from consumer products. This includes investigations of children's products, pollutant emission of chemicals and biological pollutants from structural materials, consumer products, and indoor combustion services; their impact on indoor air quality, bioavailability, and potential for consumer exposure to carcinogens and other chemical hazard-containing commercial substances; and acute and chemical toxicity of various household products to determine proper precautionary and first aid labeling. The directorate is also responsible for implementing the Poison Prevention Packaging Act with respect to household substances.

Publications: Findings are published in scientific journals and research reports awarded from National Technical Information Service.

Council of American Building Officials (CABO)
5203 Leesburg Pike, Suite 708
Falls Church, VA 22041
Phone: 703-931-4533
Fax: 703-379-1546

Description: Founded 1972. Has membership of about 8,000 states, counties, cities, and towns. Has four staffpeople. Members also include builders concerned with the building code. Holds annual meeting and board meeting.

Purpose: To promote building codes.

Activities: Includes three model code organizations. Develops, recommends, and promotes new products, uniform regulations, and new model codes. Maintains Board for the Coordination of the Model Codes and National Evaluation Service. Administers a national Certified Building Official Program.

Publications: None.

Cure Formaldehyde Poisoning Association (CURE)
9255 Lynwood Road
Waconia, MN 55387
Phone: 612-442-4665

Description: Founded 1980. Members are health professionals, lawyers, and anyone who is interested in the toxic effects of formaldehyde. Holds annual meeting.

Purpose: To educate about the problems caused by formaldehyde.

Activities: Lobbies on matters relating to formaldehyde. Conducts seminars. CURE is an acronym for Citizens United to Reduce Emissions (of Formaldehyde Poisoning Association).

Publications: Environmental Guardian, quarterly.

Hardwood, Plywood, and Veneer Association (HPVA)
1825 Michael Faraday Drive
PO Box 2789
Reston, VA 22090

Phone: 703-435-2900
Fax: 703-435-2537

Description: Founded 1921. Has 175 members and ten staffpeople. Is multinational. Budget is $1.2 million. Members are manufacturers and salespersons of plywood and veneer. Holds semiannual conference in the spring and fall.

Purpose: Promotes the manufacture and sale of plywood, veneer, and their related products. Conducts laboratory testing of plywood, adhesives, formaldehyde emissions, and smoke density. Provides public relations, advertising, marketing, and technical services to members. Represents the industry in legislative matters. Keeps members informed on tariff and trade actions. Has reference library on such subjects as plywood and veneer manufacturing and uses.

Publications: Executive Brief, quarterly; *Hardwood Plywood and Veneer News,* six times per year; *Where to Buy Hardwood Plywood and Veneer,* annual. Also publishes executive briefs, standards, and design guides.

International Conference of Building Officials (ICBO)
5360 Workman Mill Road
Whittier, CA 90601-2298
Phone: 310-699-0541
Fax: 310-692-3853

Description: Founded 1922. Has 15,000 members and 135 staffpeople. Membership dues $195 annually (government). Budget is $15 million. There are five regional groups, 88 state groups, and three local groups. Is multinational. Members are from local, regional, and state governments. Annual business meeting always in September.

Purpose: To promote the Uniform Building Code and its enforcement.

Activities: Investigates and researches safety to life and property in construction. Develops and promulgates uniformity in building construction. Educates building officials. Formulates guidelines for building inspection departments. Conducts training programs for code inspectors. Maintains speakers' bureau. Has library of 250 archival material. Bestows the A. J. Lund Award each year for outstanding contribution to the group's objectives; the John Fies Award each year for outstanding building and industry representative; and the Phil Roberts Award each year to an outstanding building official. Works through several departments.

Publications: Building Standards Magazine, monthly; *Building Standards Newsletter,* monthly; *ICBO Certification Roster,* annual; *ICBO Membership Roster,* annual; *The Uniform Building Code,* triennial. Also publishes material on dwelling house construction and building inspection, a travel manual for building inspectors, and other codes, manuals, and textbooks. Has videos on building code, enforcement, and administration available.

International Ozone Association (IOA)
c/o Pan American Group
31 Strawberry Hill Avenue
Stamford, CT 06902
Phone: 203-348-3542
Fax: 203-967-4845

Description: Founded 1973. Has 800 members and 3 staffpeople. Budget is $150,000. Has regional groups. Is multinational. Members are individuals, companies, consulting engineers, and service/support organizations that use ozone. Holds biennial congress.

Purpose: To promote proper use of ozone.

Activities: Works to advance positive applications of ozone. Furthers ozone technology through scientific and educational means. Presents a recognition award. Holds symposia and workshops.

Publications: Ozone News, bimonthly; *Ozone Science and Engineering,* bimonthly; *Who's Who in the World of Ozone;* and *Proceedings.*

National Fire Protection Authority (NFPA)
(American National Standards Institute)
1 Batterymarch Park
P.O. Box 9101
Quincy, MA 02269-9101
Phone: 617-770-3000
Fax: 617-770-0070

Description: Founded 1896. Has 65,000 members and 300 staffpeople. Budget is $40 million. There are three regional groups.

Membership includes people from fire service, business and industry, health care, educational institutions, insurance, government, architecture, and engineering. Annual meeting is held in November.

Purpose: To promote fire safety.

Activities: Develops and disseminates standards to help minimize effects of fire and explosion. Conducts fire safety education programs for the public. Provides information on fire protection and prevention. Compiles statistics on causes and occupancies of fires, loss, deaths, and firefighter casualties. Provides field service by specialists on flammable liquids and gas, electricity, and marine fire problems. Sponsors National Fire Prevention Week each October. Sponsors education campaigns featuring Sparky, the Fire Dog. Sponsors seminars on the Life Safety Code, the National Electrical Code, hotel/motel fire safety, and general fire safety. Conducts research projects. Maintains library of 20,000 books, periodicals, and audiovisual materials. Has 13,000 pieces of microform. Bestows the annual Paul C. Lamb Award for outstanding contribution to development of fire safety standards. Has many sections in the association.

Publications: Catalogs of Publications and Visual Aids, annual; *Fire News,* bimonthly; *Fire Protection Handbook; Fire Protection Reference Directory and Buyer's Guide,* annual; *Learn Not to Burn Curriculum; National Fire Codes,* annual; *National Fire Protection Association—Technical Committee Reports/Technical Committee Documentation,* semiannual; *NFPA Journal,* bimonthly; reports, and yearbook.

National Institute of Building Sciences (NIBS)
1201 Eye Street NW, Suite 400
Washington, DC 20005
Phone: 202-289-7800
Fax: 202-289-1092

Description: Founded 1976. Has 800 members and 25 staffpeople. Members are individuals and organizations; architects, contractors, real-estate professionals, builders; all levels of government interested in the building industry. Has divisions on codes, standards and regulations, information systems, and technology and land use. Holds conventions.

Purpose: To promote a favorable and coherent building regulatory environment and encourage new technology in the building industry to improve indoor air quality.

Activities: Holds seminars and periodic workshops. Presents awards annually, such as the Institute Honor Award; the Institute Member Award; and the President's Award.

Publications: Asbestos Guide Abatement Specifications, periodic; *Asbestos Operations and Maintenance Work Practices Manual,* periodic; *Building Sciences,* bimonthly; *The CCB Bulletin,* quarterly; *Construction, Criteria Base,* quarterly; *Metric in Construction,* bimonthly; and *National Institute of Building Sciences—Annual Report to the President.*

National Particleboard Association (NPA)
18928 Premiere Court
Gaithersburg, MD 20879
Phone: 301-670-0604
Fax: 301-840-1252

Description: Founded 1960. Has 18 members and 12 staffpeople. Is multinational. Has $1,648,400 budget. Members are manufacturers. Holds semiannual meeting.

Purpose: To promote manufacturing standards.

Activities: Establishes industry product standards with the American National Standards Institute for quality and performance. Sponsors educational programs, compiles statistics, conducts industry surveys, maintains a reference library (not open to public) of books, periodicals, and other materials. Presents Robert E. Dougherty Award annually. Does have committees.

Publications: None.

National Toxicology Program
P.O. Box 12233
Research Triangle Park, NC 27709

Description: Founded 1977. Its basic mission is to strengthen the science base in toxicology and coordinate research in testing activities on toxic chemicals.

Purpose and Activities: The purposes of the program are to develop a series of lists useful for public health regulation of toxic chemicals, expand the toxicological profiles of chemicals tested, increase the rate of testing of chemicals for such toxic effects as carcinogenicity, mutagenicity, and reproductive and development effects, and the dissemination of testing results of testing and methods development programs to government research and regulatory agencies, industry, labor, environmental groups, the scientific community, and the public. Chemicals chosen for testing are those that involve widespread or intense human exposure.

Publications: Research results are published in primary journals and as research reports and proceedings.

Southern Building Code Congress International (SBCCI)
900 Montclair Road
Birmingham, AL 35213
Phone: 205-591-1853
Fax: 205-592-7001

Description: Founded 1940. Has 9,200 members and 63 staffpeople. Budget is $7 million. Active members are state, county, municipal, and government officials. Associate members are trade associations, architects, engineers, and contractors. Annual meeting is held in October.

Purpose: To promote building codes and regulations for safety.

Activities: Encourages uniformity in building regulations through the Standard Codes and their enforcement. Provides technical and educational services to members and others. Provides research on new materials and methods of construction. Conducts seminars on code enforcement and inspection. Has a 350-volume reference library. Has committees on code revision.

Publications: Membership Directory, annual; *Southern Building Magazine,* bimonthly; *Standard Building Code,* triennial; *Standard Fire Prevention Code,* triennial; *Standard Gas Code,* triennial; *Standard Mechanical Code,* triennial, *Standard Plumbing Code,* triennial; pamphlets.

World Floor Covering Association (WFCA)
2211 E. Howell Avenue
Anaheim, CA 92806-6033
Phone: 800-624-6880

Description: Founded 1994. Has 1,600 members and seven staffpeople. Members are retail floor covering store owners, managers, distributors, and manufacturers. Budget is $1.2 million. There are 12 regional groups. Holds annual meeting and symposium.

Purpose: To promote retail floor covering store owners.

Activities: Provides liaison, through a Washington, D.C., lobbyist, between membership and government organizations that affect the business. Maintains Industry Education Foundation and a Group Insurance Trust. Conducts seminars to promote professional development. Compiles statistics. Conducts research. Presents Top Manufacturer Award annually.

Publications: American Floorcovering Association—Annual Management Report; American Floorcovering Association—Employer's Handbook, periodic; *Contract Primer; Family and Medical Leave Act Compliance Guide; Indoor Air Quality Brochure; Industry Issues,* quarterly; *The Square Yard,* monthly.

Sources

Encyclopedia of Associations, vol. 1: National Organizations of the United States. Detroit: Gale Research, 1998.

Encyclopedia of Governmental Advisory Organizations, 12th ed. Detroit: Gale Research, 1998.

Government Research Directory, 1995–1996. 8th ed. Detroit: Gale Research, 1994.

The United States Government Manual, 1997–1998. Washington, DC: U.S. Government Printing Office, 1997.

Bibliography 4

There has been a significant increase in the literature on indoor pollution since 1975 as a response to the recognition of the possible dangers to human health. Indoor contamination is now recognized by the public as one of the nation's major health problems. Intrinsic to this concern is not only the removal of devastating events from the past but the need to provide a quality indoor environment in the future. The literature varies from popular accounts to the most technical scientific papers. The selection of bibliographic material provides a wide perspective of modern indoor contaminants. The final section of this chapter lists selected journals that publish articles on indoor pollution.

Reference Sources

Air Pollution

The Indoor Air Quality Directory, 1994–1995: Residential, Commercial, Industrial. Chevy Chase, MD: IAQ Publications, 1994, 378 pp. ISBN 0-9633003-5-0.

Leininger, A., et al. *Catalog of Materials as Sources of Potential Indoor Air Emissions, Project Summary.* 94-0483-M (vol. 1). Research Triangle Park, NC:

U.S. Environmental Protection Agency, Air and Research Laboratory, 1993. Microfiche.

Plog, Barbara A., George S. Benjamin, and Maureen A. Kerwin. *Fundamentals of Industrial Hygiene.* Chicago: National Safety Council, 1988.

Skeist, I. *Handbook of Adhesives.* 3d ed. New York: Van Nostrand, 1990.

U.S. Environmental Protection Agency. United States Public Health Services. National Environmental Health Association. *Introduction to Indoor Air Quality: A Manual.* EPA/400/3-91/003. Denver: National Environmental Health Association, 1991, 297 pp. Microfiche.

U.S. National Institute for Occupational Safety and Health. Division of Standards Development and Technology Transfer. *Indoor Air Quality: Selected References.* Cincinnati: 1989, 60 pp.

Air Quality

Baca, David R. *Asbestos: A Selected Bibliography of Journal Articles.* Public Administration Series: Bibl. no. P2751. Monticello, IL: Vance Bibliographies, 1989, 8 pp.

U.S. Department of Health and Human Services. Office on Smoking and Health. Technical Information Center. *Bibliography on Smoking and Health, 1987.* Public Health Service Bibliography Series no. 45. Rockville, MD: 1988, 591 pp.

———. *Bibliography on Smoking and Health, 1985.* Public Health Service Bibliography Series no. 45. Rockville, MD: 1986, 552 pp.

U.S. Environmental Protection Agency. Office of Health and Environmental Assessment. Environmental Criteria and Assessment Office. *Indoor Air: Reference Bibliography.* Research Triangle Park, NC: 1989, 274 pp.

Younger, Gregory. *Indoor Radon: A Selected and Annotated Bibliography.* CPL Bibliography no. 184. Chicago: Council of Planning Librarians, 1987, 26 pp.

General

Baird, John C., Birgitta Berglund, and William T. Jackson, eds. *Indoor Air Quality for People and Plants.* Stockholm: Swedish Council for Building Research, 1991, 188 pp. ISBN 91-540-5263-7.

Material for this book was obtained from a conference held in 1987 at Dartmouth College and sponsored by the Rockefeller Center for the Social Sciences. A group of psychologists, public health researchers, and botanists met to study the potential for using plants and biological species to indicate if the indoor air of a building is healthy or sick. The first part of the book deals with air quality in sick buildings; the second part deals with bioassays of air quality in nonhuman systems. More than 300 organic compounds have been found in indoor air that are caused by building materials, combustion processes, room furnishings, paints, carpet cleaners, consumer products, and outdoor air. Chapters have bibliographic references, some tables, and figures.

Boubel, Richard W., Donald L. Fox, D. Bruce Turner, and Arthur C. Stern. *Fundamentals of Air Pollution.* 3d ed. San Diego: Academic Press, 1994, 574 pp. ISBN 0-12-118930-9.

This is an excellent book to read that tells the history and the effects of air pollution. One section offers ways to monitor and measure air pollution, including regulatory and engineering control of air pollution. A special chapter deals with indoor air quality, discussing indoor air pollutants and their effects on humans. Methods to control indoor pollutants such as molds, toxins, and radon are discussed. Biological contaminants were recognized in 1976 with the outbreak of Legionnaires' disease in Philadelphia, resulting in many deaths. The contaminant was found in the hotel's air-conditioning system. Sick building syndrome is also discussed. Environmental tobacco smoke (ETS) control is quite complicated, because it deals with the behavior of individuals. Control of biological agents depends on cleaning and maintenance of heating and air-conditioning systems. Volatile organic compounds include building materials, wall and floor coverings, and furnishings and needs to be controlled by the manufacturer and builder. This book has special tables on indoor air pollutants and a few bibliographic references.

Coffel, Steve, and Karyn Feiden. *Indoor Pollution.* New York: Fawcett Columbine, 1991, 278 pp. ISBN 0-449-90476-8.

The book begins by giving information about the sources of indoor air pollution and water pollution and health hazards, offering ways to reduce and even eliminate them. Several chapters are devoted to sick building syndrome–related problems and how worker health is affected. The final section recommends ways that federal, state, and local governments can get involved to help solve the problems of indoor air pollution. Many tables and diagrams illustrate the book. Included in the appendices are a glossary, a bibliography, a checklist for finding pollution problems, organizations involved with indoor pollution, testing equipment, and laboratories, and federal air and water legislation. This book is very easy reading and understandable.

DuPont, Peter, and John Morrill. *Residential Indoor Air Quality and Energy Efficiency.* Series on Conservation and Energy Policy. Berkeley and Washington, DC: American Council for an Energy-Efficient Economy in cooperation with National Rural Electric Cooperation Association, Washington, DC, and Universitywide Energy Research Group, University of California, 1989, 267 pp. ISBN 0-918249-08-2.

For years the connection between indoor air quality and energy efficiency has been overlooked. Homes have been tightened to save energy and as a result pollutants have been trapped, thereby affecting the quality of indoor air. In order to prevent this it is necessary to reduce or eliminate the sources of pollutants. This volume examines pollutants to see which are most harmful and how to eliminate them. The ventilation needed for healthy air is discussed. The book should be of special interest to those in the building, conservation, and health fields. The book is illustrated with many figures and tables. There is a glossary, and references are found for each chapter at pp. 239–256.

Godish, Thad. *Air Quality.* Chelsea, MI: Lewis Publishers, 1985, 372 pp. ISBN 0-87371-019-3.

This book provides a comprehensive overview of air pollution, with primary emphasis on control technologies and quantitative considerations. Although much of the book considers the problems of atmospheric pollutants, a major section is devoted to indoor pollution. Topics include personal air pollution exposure, sick building syndrome, combustion by-products, asbestos, radon, formaldehyde, organic chemicals, pesticides, biogenic pollutants, abatement measures, role of energy conservation, and implications for public policy. This volume is not directed to a particular discipline. It provides an excellent background to more specialized material. The book has many tables and figures. Each chapter has a list of references, readings, and questions; includes index.

Maroni, Marco, Bernd Seifert, and Thomas Lindvall, eds. *Indoor Air Quality: A Comprehensive Reference Book.* Air Quality Monographs, vol. 3. New York: Elsevier, 1995, 1049 pp. ISBN 0-744-81642-9.

This book was written to show that to achieve, maintain, and restore indoor air quality it requires the work of architects, engineers, hygienists, doctors, biologists, chemists, environmentalists, and many others. The book is divided into eight parts. Part 1 discusses the nature, sources, and toxicity of indoor air pollutants. Part 2 talks about the health effects of indoor air pollution. Part 3 discusses risk assessment. Part 4 investigates and diagnoses illnesses and complaints related to buildings. Part 5 considers the dynamics of indoor air contaminants. Part 6 investigates indoor air quality in buildings. Part 7 looks into the control of indoor air quality and climate. Part 8 presents guidelines for indoor air quality. References follow each part; many tables and figures; includes an excellent analytical index.

Meckler, Milton, ed. *Indoor Air Quality Design Guidebook.* Lilburn, GA: Fairmont Press, 1991, 283 pp. ISBN 0-88173-088-2.

Several authors contributed to this volume. The first part is devoted to sources of indoor air pollution and health effects. Such topics as formaldehyde, radon, particulates, unventilated space heaters, wood-burning stoves and fireplaces, tobacco smoke, different household products, and volatile organic compounds are discussed at length. The final chapter in part 1 deals with sick building syndrome and effects on human health. Part 2 is devoted to solutions to indoor air quality problems. Figures and tables illustrate the book. References are given for each chapter.

Meyer, Beat. *Indoor Air Quality.* Reading, MA: Addison-Wesley, 1982, 434 pp. ISBN 0-201-05094-3.

The goal of this book is to describe some of the factors that determine indoor air quality, to review the status of current knowledge, and to provide an updated list of publications that describe the frontiers of research. The introductory chapter gives an overview of the indoor air problem and is followed by chapters on history of development, parameters that define the health and welfare of building occupants, selected building factors, prevailing air pollutants and their sources, current air monitoring and analysis methods, knowledge of pollutant concentrations, and the complex field of health effects; reviews of the extensive control literature and the U.S. legislative and voluntary regulating controls are also provided. An author index and a subject index are also included; figures and tables illustrate the book. Appendix 1 contains units and conversion factors used in the book. Appendix 2 lists acronyms and abbreviations used in the book. References and bibliography are found on pp. 337–397.

Moffat, Donald W. *Handbook of Indoor Air Quality Management.* Englewood Cliffs, NJ: Prentice Hall, 1997, 506 pp. ISBN 0-13-235300-0.

This large volume is divided into five parts. Part 1 discusses what clean air means to people. It classifies undesirable elements, their sources, and health effects of contaminants. Part 2 discusses reasons for having clean air, respirators, how temperature and humidity affect workers, and effects of aldehydes on indoor air. Part 3 monitors and analyzes air samples to show their effect on people. Part 4 has many good chapters discussing ways to maintain indoor air quality. Part 5 contains five chapters that discuss ways to handle emergencies. The appendixes offer case studies, exposure limits for air contaminants, a medical reference chart, glossary, a list of agencies and associations, and several pages of forms and checklists.

Sterling, E., C. Bieva, and C. Collett, eds. *Building Design, Technology, and Occupant Well-Being in Temperate Climates.* Atlanta: American Society of Heating, Refrigerating, and Air-Conditioning Engineers, 1993, 411 pp. ISBN 1-883413-05-2.

This volume is a compilation of papers given at the International Conference on Building Design, Technology, and Occupant Well-Being held in Brussels, Belgium, February 17–19, 1993. The main topics are design, operation, and maintenance; energy management; environmental management; building performance and thermal comfort assessment; indoor environmental quality; occupant health and well-being; and legislation, standards, and litigation. Papers are illustrated with tables and figures. Each paper contains an abstract and references.

Wadden, Richard A., and Peter Scheff. *Indoor Air Pollution: Characterization, Prediction, and Control.* Environmental Science and Technology Series. A Wiley-Interscience Publication. New York: John Wiley and Sons, 1982, 213 pp. ISBN 0-471-87673-9.

The purpose of this book is to review current information about indoor pollution hazards and to discuss the tools available for characterizing not only the problem but also some solutions. The major thrust of the volume is to provide a heightened awareness of the considerations that need to be taken into account to define the problems and provide solutions. The book is organized into four parts: characterization, prediction, control, and application. The book takes a general approach in order to provide the users some appreciation of how background information can be integrated into a realistic evolution of the indoor pollution problem. This volume was originally written for a course offered under the auspices of the Air Pollution Control Association. The book has many tables and an index. References are found with each chapter.

Walsh, Phillip J., Charles S. Dudney, and Emily D. Copenhaver, eds. *Indoor Air Quality.* Boca Raton, FL: CRC Press, 1984, 207 pp. ISBN 0-8493-5015-8.

Many authors contributed to this book. It addresses how individuals are exposed to pollutants in indoor air pollution. Such pollutants include radioactive radon gas from the ground and from building materials, formaldehyde from building materials and furniture, smoke from cigarettes, and fibers from household furnishings and everyday activities in the household. Exposure occurs long before any effects become evident. Since people spend 58–78 percent of their time indoors, the quality of indoor air is of great importance. Secondhand smoke has adverse effects on health and should be avoided. The book has more than 200 bibliographic references. There are also references for each chapter. The book is well illustrated with tables and figures.

Warde, John. *The Healthy Home Handbook: All You Need to Know to Rid Your Home of Health and Safety Hazards.* New York: Random House, 1997, 388 pp. ISBN 0-8129-2151-8.

This book provides practical instructions for eliminating household pollutants, poison, and safety hazards. Part 1, "Controlling Indoor Pollutants," contains individual chapters covering all of the harmful substances that can invade a house, such as radon, biological contaminants, and volatile organic compounds. Part 2 covers causes of physical injury in houses. These include protection against fires and falls, the two most common causes of household accidents, electrical problems, childproofing, and providing comfortable accommodations for the elderly and disabled.

The book concludes with appendices and with a discussion of the latest findings on three controversial topics: electromagnetic fields, multiple chemical sensitivity, and seasonal affective disorder.

The final appendix contains a selected list of books, organizations, laboratories, and companies where help can be obtained to solve the problem of indoor pollutants. The book is very well illustrated—a list giving the page on which the illustration is found at pp. xvii–xxiii.

Indoor Pollution and Health

Building Performance and Regulations Committee/Committee on the Environment. *Designing Healthy Buildings: Indoor Air Quality.* Washington, DC: American Institute of Architects, 1993, 173 pp.

A panel of six judges (two from Committee on the Environment, two from Building Performance and Regulations, one from Interior, and one international expert) chose papers presented at a symposium held in November 1992 in Los Angeles, California. The purpose of the meeting was to exchange and disseminate information about indoor air quality. Papers were chosen that addressed codes and regulations, new construction, public policy, materials, renovation, and overall approach to quality indoor air. References are found with most papers; a few tables and diagrams.

Grammage, Richard B., Stephen V. Kaye, and Vivian A. Jacobs. *Indoor Air and Human Health.* Chelsea, MI: Lewis Publishers, 1985, 421 pp. ISBN 0-87371-006-1.

This volume contains the proceedings of the Seventh Life Sciences Symposium, sponsored by Oak Ridge National Laboratory and others, which was held at Knoxville, Tennessee, on October 29–31, 1985. At this meeting papers were given on such topics as radon, microorganisms, passive cigarette smoke, organics, and combustion products. Papers are well illustrated with figures and tables and contain bibliographic references. The papers on passive cigarette smoke are very informative.

Johnson, Bertel G., Johnny Kronvall, Thomas Lindvall, Allan Wallin, and Hanne Weis Lindencrona. *Buildings and Health: Indoor Climate and Effective Energy Use.* Stockholm: Swedish Council for Building Research, 1991, 176 pp. ISBN 91-540-5369-X.

Health problems associated with indoor climate seem to be on the increase in Sweden and are related to energy conservation and new energy technology. Part 1 is devoted to the human being and the indoor climate. Part 2 discusses indoor climate and techniques such as energy, heating, cooling, ventilation, lighting, and noise sources. Part 3 examines ways to improve the indoor climate through building design, physical planning, and climate requirements. Figures and tables are found throughout the book. Sources and literature are found within chapters. Key words, concepts, and terms are listed at the end of book. There is no index, so the table of contents is invaluable in using this volume.

Seifert, B., H. J. van de Wiel, B. Dodet, and I. K. O'Neill, eds. *Environmental Carcinogens Methods of Analysis and Exposure Measurement: Indoor Air.* Vol. 12. IARC Scientific Publications No. 109. Lyon, France: International Agency for Research on Cancer, 1993, 384 pp. ISBN 92-832-2109-5.

Humans are exposed to toxic materials in buildings where they eat, sleep, work, and relax. Heretofore, contaminants have focused on tobacco and other smokes because of their noxious properties. It is now a common problem because the dangers of radon, sick building syndrome, cooking fumes, and synthetic building and furniture materials have become better known. This volume analyzes known carcinogens and shows effects on humans. Tables and figures are illustrative; references are found throughout. Several pages list publications of the International Agency for Research on Cancer (IARC) and the IARC Monograms on the evaluation of carcinogenic risks to humans.

Smith, Kirk R. *Biofuels, Air Pollution, and Health: A Global Review.* Modern Perspectives in Energy Science. New York: Plenum Press, 1987, 452 pp. ISBN 0-306-42519-X.

This book emphasizes the fact that childhood respiratory diseases and chronic lung disorders are caused by exposure to emission from biomass cooking and heating fuels, especially in developing countries. The book is a result of the work of the Biofuels and Development Project of the East-West Center. Such pollutants as carbon monoxide, formaldehyde, and cigarette smoke are discussed. The book contains five appendices and many references and is well illustrated with tables and figures. A list of tables is included in the front matter of the book.

Tate, Nicholas. *The Sick Building Syndrome: How Indoor Pollution Is Poisoning Your Life—And What You Can Do.* Far Hills, NJ: New Horizon Press, 1994, 182 pp. ISBN 0-88282-082-6.

This book considers the causes of sick building syndrome and examines its prevalence, costs, and health consequences; it covers residential buildings, schools, offices, and industrial workplaces. This is a comprehensive study of a modern problem created by energy conservation (which results in more airtight buildings and thus indoor pollution). There are references, tables, and figures.

Wang, Rhoda. *Water Contamination and Health: Integration of Exposure Assessment, Toxicology, and Risk Assessment.* Environmental Science and Pollution Control Series. New York: Mercel Dekker, 1994, 544 pp. ISBN 0-8247-8922-9.

This book addresses the various concerns and the pros and cons of exposure to water contaminants for health risk assessment. The unique feature of this book is the integration of exposure assessment, toxicology, and risk assessment in a single volume. The diversification of techniques applied in this book is condensed into four sections. The first section, "Water Contaminants: Monitory, Treatment, and Health Impact," covers EPA and California's efforts in water contamination. The second section, "Exposure

Assessment of Water Contaminants: Specific Dosimetry of Radon and Chloroform," emphasizes exposure assessment and dose. The next section, "Methodology Development in Exposure Assessment and Dose Estimates," discusses the most up-to-date techniques. The last section, "Human Health Risk Assessment and Water Contaminants and Toxicological End Points," deals strictly with modern risk assessment techniques applicable to water contaminants. Each chapter was written by an authority in the field. Book has many figures and tables. References are found within each chapter.

Indoor Pollution—Control

Fisk, W. J., R. K. Spencer, D. T. Grimsrud, F. J. Offermann, B. Pedersen, and R. Sextro. *Indoor Air Quality Control Techniques: Radon, Formaldehyde, Combustion Products.* Park Ridge, NJ: Noyes Data Corporation, 1987, 245 pp. ISBN 0-8155-1129-9.

In this volume the authors review and evaluate indoor air quality control techniques. Some of the pollutants addressed are radon, formaldehyde, and combustion products such as carbon monoxide, nitrogen dioxide, carbon dioxide, and respirable particles. Heretofore, attention was given to outdoor air contamination, but recently attention has turned to indoor contamination because people spend so much time indoors. Buildings have various sources of indoor air pollution: people, pets, carpets, furniture, building materials, tobacco, and the soil under and around the building. Sources must be understood in order to control them. Bibliographic references are found at the end of the book. Book also has tables and figures for illustration.

Godish, Thad. *Indoor Air Pollution Control.* Chelsea, MI: Lewis Publishers, 1989, 401 pp. ISBN 0-87371-098-3.

Because indoor air pollution is a relatively recent concern in public health, there is a need for reference books to provide basic information. This volume serves that need. It contains an overview of the problem and expanded discussion of source control measures for specific contaminants. It also reviews public policy and regulatory issues associated with the problem.

The author expands the utility of the book by including chapters on problem solving, including air quality diagnosis of specific indoor air contamination problems in residences and public buildings, mitigation practices, and practical guides to identifying and solving real-world problems. These chapters complement the theory and principles, the primary focus of the reference portion of the book.

This book is intended for a wide audience, including public health and environmental professionals, industrial hygienists, architects, physicians, and academia. Illustrated with tables and figures; references are found within each chapter. An index provides access to items discussed in the book.

Hansen, Shirley J. *Managing Indoor Air Quality.* Lilburn, GA: Fairmont Press, 1991, 117 pp. ISBN 0-88173-107-2.

Indoor air quality has been a problem for years, and it isn't going away. IAQ even in "healthy" buildings must be considered. This nontechnical book aims to help those responsible for managing the environment, for it encompasses the various contaminants, molds, fungi, toxic gases, thermal conditions, and any factors that make air uncomfortable in buildings. The book serves as a guide and reference to help prevent IAQ problems from occurring and treat those already existing. An IAQ consultant should always be consulted if a building shows signs of sick building syndrome. There is a list of selected resources and references. The book contains a glossary of terms and a glossary of acronyms and abbreviations. Appendix A addresses contaminants; appendix B deals with sample contaminant protocol; appendix C shows investigation forms. Illustrated with many figures.

Harp, Steve M., Ronald V. Gobbell, and Nicholas R. Ganick. *Indoor Air Quality: Solutions and Strategies.* New York: McGraw Hill, 1995, 417 pp. ISBN 0-07-027373-1.

This book was written by an architect, an industrial hygienist with a chemical engineering background, and a mechanical engineer. The purpose is to provide a foundation of IAQ analysis and control. Indoor air quality is a complex problem because it is affected by hundreds of pollutants, and the indoor environment must be healthy if it is to be comfortable for occupants. Sick building syndrome is common in buildings and is a building owner's nightmare. Building related illness is another condition associated with unhealthy buildings. A lengthy chapter is devoted to IAQ symptoms, SBS, and BRI. Another chapter is devoted to show how heating, ventilation, and air-conditioning can affect the air quality. One chapter shows how architecture, construction, and operations affect the quality of air in buildings. A short chapter summarizes the book. There are five appendices, a glossary, and a list of symbols and abbreviations used in the book. References are found at the end of each chapter; illustrated with figures and tables.

Hines, Anthony L., Tushar K. Ghosh, Sudarshon K. Loyalka, and Richard C. Warder Jr. *Indoor Air: Quality and Control.* Englewood Cliffs, NJ: PIR Prentice Hall, 1993, 340 pp. ISBN 0-13-463977-4.

This volume was written to help people better understand indoor air pollution problems and ways to solve them. Such topics as volatile organic pollutants, inorganic gaseous pollutants, heavy metals, respirable particulates, bioaerosols, and radon are discussed with methods to help solve the problems. The last two chapters involve adsorption/absorption processes for indoor air pollution control. There are five appendixes, a subject index, an and author index. Chapters contain tables for illustration and bibliographic references.

Meckler, Milton. *Improving Indoor Air Quality Through Design, Operation, and Maintenance.* Lilburn, GA: Fairmont Press, 1996, 272 pp. ISBN 0-88173-208-7.

Concentrations of harmful contaminants are higher indoors than they are outdoors and pose a threat to human health. Indoor air quality is affected by contamination in a building, ventilation, and dilution of indoor contaminants with outside air. IAQ is determined by building materials, cleaners, furnishings, combustion appliances and processes, and occupants. Since we spend so much time indoors, contamination sources have an effect on human health and comfort. Heating, ventilation, and air-conditioning are responsible for the quality of indoor air and need to be operated and maintained properly. References are found at the end of each chapter. The book is well illustrated with tables and figures.

Meckler, Milton, ed. *Indoor Air Quality Design Guidebook.* Lilburn, GA: Fairmont Press, 1990, 283 pp. ISBN 0-88173-088-2.

The primary objective of this book is to outline major indoor air pollutants and provide engineering solutions to indoor air quality problems. The volume begins with a discussion of IAQ problems such as formaldehyde, radon, particulates, carbon monoxide, and others. These chapters are followed by engineering solutions on control of indoor air pollution, evaluation of methods for measuring major indoor air pollutants, techniques for modeling ventilation efficiency, indoor air quality simulation by using computer models, zoning for indoor air quality, and solid desiccants and air quality. The volume concludes with a discussion of systems design and maintenance guidelines. The *Indoor Air Quality Design Guidebook* is a consolidation of air problems and controls for use by architects, engineers, contractors, homeowners, building managers, and building operators who want to provide a healthy environment for all types of building occupants. Figures and tables illustrate the book. Bibliographic references are found at the ends of chapters. An index is included.

Moffat, Donald W. *Handbook of Indoor Air Quality Management.* Englewood Cliffs, NJ: Prentice Hall, 1997, 506 pp. ISBN 0-13-235300-8.

This large volume, dealing with indoor air quality for industry and workplaces, is divided into five parts. Part 1 gives an overview of indoor air pollution. Part 2 gives regulations and how temperature and humidity affect workers. Part 3 discusses monitoring indoor air quality. Part 4 examines ways to maintain indoor air quality. Part 5 explains ways to manage emergencies. Illustrated with tables and figures. In the appendices the author gives case studies, medical reference charts, glossary, exposures limits for air contaminants, a list of agencies and associations, and examples of forms and checklists.

Morawska, Lidia, N. D. Bofinger, and Marco Maroni, eds. *Indoor Air: An Integrated Approach.* New York: Elsevier Science, 1995, 515 pp. ISBN 0-08-041917-8.

This book is a selection of working session reports and papers presented at the International Workshop: Indoor Air-An Integrated Approach, held at the Gold Coast, Australia, on November 27–December 1, 1994. The purpose of the conference was to study the character of indoor air, develop a framework for integrated health risk assessment, develop strategies for controlling all indoor air pollutants, and look at areas for future research for overall improvement in indoor air quality. Book is illustrated with many figures and tables. There is a keyword index and an author index. Biographical sketches are given for editors; bibliographic references are used in most papers.

Nagda, Niren L., Harry E. Rector, and Michael D. Koontz. *Guidelines for Monitoring Indoor Air Quality.* New York: Hemisphere Publishing Corporation, 1987, 270 pp. ISBN 0-89116-385-9.

This book was written because of great concern over health issues related to air quality in homes, schools, offices, and buildings. The authors provide an approach for monitoring indoor air quality. The book is useful to those in government, the private sector, and academia. It begins by giving the historical background of indoor air pollution and the many factors that influence air quality. Instruments and methods for monitoring indoor air quality are discussed fully. Steps to develop a detailed design for indoor air quality monitoring are described. One chapter lists additional reading materials. Researchers in the field of indoor air quality will find this book useful as a supplementary source. There are three appendices; bibliographic references included.

Seifert, B., H. J. Van de Wiel, B. Dodet, and I. K. O'Neill, eds. *Environmental Carcinogens Methods of Analyses and Exposure Measurement.* Vol. 12. IARC Scientific Publications No. 109. Lyon, France: World Health Organization. International Agency for Research on Cancer, 1993, 384 pp. ISBN 92-832-2109-5.

Only recently have people become concerned over the quality of air in enclosed spaces. Complaints of such symptoms as irritation or dryness of mucous membranes, headaches, burning eyes, and fatigue gave cause for much concern. Study was made to determine specific pollutants such as formaldehyde and chemical compounds found in rooms. The volume is divided into two parts. The first part discusses sources, concentrations, controls, health effects, and methods of analysis for indoor pollutants. The second part contains useful methods for determining air pollutants in nonindustrial indoor environments. Book has many figures and tables for illustration; bibliographic references included. A list of IARC Publications is found in the back of the book.

Young, S., et al. *Stirring Up Innovation: Environmental Improvements in Paints and Adhesives.* New York: INFORM, 1994, 116 pp. ISBN 0-918780-63-2.

Examines the health problems associated with the use of paints and adhesives. Includes case studies of current efforts to ensure product safety by U.S. companies. There are references and a few tables and figures.

Laws and Regulations

Cross, Frank B. *Legal Responses to Indoor Air Pollution.* Westport, CT: Quorum Books, 1990, 206 pp. ISBN 0-89930-519-9.

Many think of air pollution as being caused only by heavy industries or automobile traffic but don't realize that indoor pollution is more hazardous. Indoor pollution applies to the home, office, restaurants, and other areas within buildings. Part 1 summarizes the health risks presented by indoor pollution, especially radon, asbestos, and formaldehyde. Part 2 discusses what federal and state governments can do to help solve the problem. Part 3 discusses ways someone can sue for health problems caused by contaminated indoor air. The last part suggests ways to protect the health of people against indoor air pollution, as it is the greatest involuntary environmental risk to human health. Most chapters are illustrated with exhibits. Notes are found at the end of each chapter; selected bibliography and list of exhibits included.

Carbon Monoxide

Shephard, Roy J. *Carbon Monoxide: The Silent Killer.* The Bannerstone Division of American Lectures in Environmental Studies. Springfield, IL: Charles C. Thomas, 1983, 220 pp. ISBN 0-398-04850-9.

This interesting volume discusses carbon monoxide yesterday and today—its sources and properties. One chapter addresses the biology of carbon monoxide and health effects of exposure to carbon monoxide. Another points out those people most vulnerable to carbon monoxide and possible acclimatization. One section looks into the accumulation of carbon monoxide in the indoor environment, with tables showing levels and concentrations. Book contains several tables and several pages of references.

Asbestos

Cherry, Kenneth F. *Asbestos: Engineering Management and Control.* Chelsea, MI: Lewis Publishers, 1988, 265 pp. ISBN 0-87371-127-0.

At one time asbestos was considered to be "nature's wonder fiber," but today it is considered to be dangerous. Its disease-causing properties have meant large expenditures in private and public areas. To protect us from the danger, government agencies have enacted rules, regulations, and guidelines. In this book, we find the viewpoints of EPA, OSHA, and the National Asbestos Council. There are four appendixes: one lists asbestos substitutes; one gives sources for asbestos information; another gives Asbestos Hazard Emergency Response Act training requirements; the last lists costs of typical equipment such as gloves, coveralls, boots, tools, wastebags, air equipment, and filters. There are a few tables and figures illustrating the book. There are a few bibliographic references.

Minerals Yearbook: Asbestos. Pittsburgh: U.S. Department of the Interior, Bureau of Mines, Yearly.

Radon

Brookins, Douglas G. *The Indoor Radon Problem.* New York: Columbia University Press, 1990, 229 pp. ISBN 0-231-06748-8.

The health risk of radon that accumulates in buildings under normal living conditions has only recently become significant. Radon from soil, building materials, water, and air causes many lung cancer deaths. Buildings containing radon can be remediated, usually with good results. Indoor radon is colorless and tasteless and can be detected only by chemical or radiation measuring techniques. Radon is radioactive and decays with a half-life of 3.8 days. But the real problem is caused by radon daughters we inhale along with radon, which is discussed in chapter 4. The purpose of this book is to educate people about radon. The book is easily read, with many figures and tables used in explanation. An appendix gives addresses for state agencies with radon information. There is a glossary and bibliographic references.

Edelstein, Michael R., and William J. Makofske. *Radon's Deadly Daughters: Science, Environmental Policy, and the Politics of Risk.* Lanham, MD: Rowman and Littlefield, 1997, 320 pp. ISBN 0-8476-8333-8.

Although the dangers from radioactive contamination became evident to the general public with the catastrophes of Three

Mile Island and Chernobyl, the danger from radon in homes is still largely ignored. The authors illustrate how scientific risk factors are ignored. In addition, political and economic influences dominating environmental policy can seriously undermine an effective response. The volume is written for the general public and shows that the hazards of radon have great relevance as people are increasingly surrounded by invisible environmental contaminants. Under these conditions, individuals can choose apathetic acceptance or attempt to discover how to actively protect themselves.

Kay, Jack G., George E. Keller, and Jay F. Miller. *Indoor Air Pollution: Radon, Bioaerosols, and VOCs.* Chelsea, MI: Lewis Publishers, 1991, 259 pp. ISBN 0-87371-309-5.

The authors wrote this book following a symposium entitled, "Air Pollution: Its Causes, Its Measurement, and Possible Solutions," presented at the 198th National Meeting of the American Chemical Society, Miami Beach, Florida, September 10–13, 1989. Papers were presented on such aspects of indoor air pollution as liability for indoor air pollution; biological pollutants; abatement of microbial contamination; indoor ozone exposures; "building bake-out" conditions on volatile organic compound emissions; and air cleaners for indoor air pollution control. The second part of the book is devoted to radon, the behavior of radon decay, and controlling radon in the indoor environment. An appendix lists people who contributed to the book. Chapters are well illustrated with figures and tables. References accompany each chapter.

Lao, Kenneth Q. *Controlling Indoor Radon: Measurement, Mitigation, and Prevention.* New York: Van Nostrand, 1990, 272 pp. ISBN 0-442-23754-5.

Radon has become a household word within recent years. The health risks posed by indoor radon have become a growing concern. This book covers such issues as physical and chemical properties of radon, biological effects of radiation, testing procedures for indoor radon, risk factors, prevention techniques, radon in water, and legal issues involving radon. Book has six appendixes, a glossary, and figures and tables. Each chapter ends with references.

Majundar, Shyamal K., Robert F. Schmalz, and E. Willard Miller, eds. *Environmental Radon: Occurrence, Control, and Health Hazards.*

Easton: Pennsylvania Academy of Science, 1990, 435 pp. ISBN 0-745809-03-4.

The health hazards of radon became apparent only in the 1980s, when it was recognized that radon contamination of buildings was widely distributed over the earth. The seriousness of long-term exposure to radon cannot be overstated. The great difficulty of alerting the general public to the devastating effects of radon is that it requires years of exposure before a health problem develops. Then it is too late. The objective of this volume is not only to present a review of the occurrence and properties of radon but to provide insights into the effects on human health. This volume is divided into six parts, beginning with a historical perspective. The second part deals with the geologic aspects of radon. The third part addresses detection and measurement of radon. Part 4 is devoted to radon health problems, with emphasis on lung dosimetry and cancer. The final part considers the regulation and policies of radon control, legal aspects, and economic impacts. This book is not only for experts who are evaluating the radon problem but will be of value to the wide audience that is just now awakening to the perils of radon contamination.

Mueller Associates, Syscon Corporation, and Brookhaven National Laboratory. *Handbook of Radon in Buildings: Detection, Safety, and Control.* New York: Hemisphere Publishing Corporation, 1988, 261 pp. ISBN 0-89116-823-0.

This book was completed under the direction and authority of the U.S. Department of Energy's Office of Environmental Analysis. Radon as an indoor air pollutant has increased because homes have been built with very little ventilation in order to save energy. The handbook is divided into seven sections. Section 1 gives a summary of issues associated with radon and its decay products on indoor air quality and health effects. Section 2 discusses sources and transport mechanisms. Section 3 examines factors that influence indoor concentrations. Sections 4 and 5 provide information on health effects. Sections 6 and 7 give models for the control of radon. Two appendices deal with radon in the home and are intended as a primer for homeowners. Book has tables and figures, a glossary, and bibliographic references at the ends of chapters.

Passive Tobacco Smoke

Ecobichon, Donald J., and Joseph M. Wie. *Environmental Tobacco Smoke.* Proceedings of the International Symposium at McGill University, 1989. Lexington, MA: Lexington Books, 1990, 389 pp. ISBN 0-669-24365-5.

This volume contains the proceedings of a meeting held at McGill University in 1989. Part 1 discusses the characteristics of environmental tobacco smoke (ETS). Part 2 assesses the exposure to ETS. Part 3 discusses tobacco smoke and the various types of cancer it causes. Part 4 discusses tobacco smoke and cardiovascular disease. Part 5 addresses effects of tobacco smoke on pulmonary function and respiratory health. Part 6 explains effects of smoking on prenatal development. Part 7 addresses risk assessments relating to environmental smoking. Part 8 has an excellent discussion on sources and the effects of smoking on indoor pollution. A few tables and figures illustrate the book. References can be found at the ends of some sections. An appendix gives the agenda. Participants are listed in the front matter.

Tollison, Robert D., ed. *Clearing the Air: Perspectives on Environmental Tobacco Smoke.* Lexington, MA: Lexington Books, 1988, 148 pp. ISBN 0-669-18007-6.

The purpose of this book is to point out views of special individuals on environmental tobacco smoke. The first few chapters, discussing health aspects, explain what indoor air is and how big a problem it poses in comparison to other pollutants. To clean indoor air requires better ventilation systems. One section of the book addresses the economic aspects. The economist feels that indoor air is owned by the restaurant owner, the factory owner, and so on; thus proper indoor air environment must be provided. After reading the book, one feels that smokers and nonsmokers will work out their differences; that labor unions will bargain with management about smoking on the job; that bars and restaurants will solve the smoking problem; that problems concerning ETS will be solved. An interesting appendix contains quotations showing that normal levels of ETS do not necessarily pose health problems. Information is given about contributors. Chapters have a few notes and bibliographic references.

Smoking Hazards

Committee on Advances in Assessing Human Exposure to Airborne Pollutants. *Human Exposure Assessment to Airborne Pollutants: Advances and Opportunities.* Washington, DC: National Academy of Sciences, 1991, 321 pp. ISBN 0-309-04284-4.

As part of the attempts to better understand human exposures to hazardous substances, the Agency for Toxic Substances and Disease Registry (ATSDR) sponsored this study of advances in assessing exposure to airborne contaminants. Numerous techniques have evolved concurrently to qualitatively and quantitatively establish numerous profiles. EPA has provided a starting point to consider priorities for environmental contaminant assessment. In industrial hygiene practice, assessment of workers' exposure to pollutants has been conducted for many years. From these evaluations, the techniques of industrial hygiene are now being refined and, with other new techniques, are being applied to the industrial community. This book considers such aspects as: frameworks for assessing exposures to air contaminants, sampling and physical-chemical measurements, research measures and exposure assessments, and models. The volume concludes with anticipated applications to VOCs, tobacco smoke, lead, asbestos, radon, toxic releases, and sick building syndrome. References are found on pp. 259–309. Book has a few tables and figures for illustration. There is a glossary and three appendices.

Douville, Judith A. *Active and Passive Smoking Hazards in the Workplace.* New York: Van Nostrand Reinhold, 1990, 221. pp. ISBN 0-442-00167-3.

This nonmedical book is designed to assist industrial, commercial, and governmental personnel directors, managers, and public officials in coping with the increasing problem posed by workplace smoking. The book provides a practical guide, background information, and advice on the hazards of workplace smoking and passive exposure to ETS. Guidance is given to justify enforcement of workplace smoking bans, including discussion of and possible solutions to technical problems encountered while banning workplace smoking. The material presented includes the hazards of active and passive smoking, employer considerations, workplace restriction, costs and benefits of the program, and program implementation and outcome. There are tables and bibliographic references. There is also a list of selected references.

Richmond, Robyn, ed. *Interventions for Smokers: An International Perspective.* Baltimore: Williams and Wilkins, 1994, 363 pp. ISBN 0-683-07272-2.

This volume is divided into three parts. Part 1 discusses individual and chemical approaches to overcome smoking. Part 2 discusses smoking cessation in specific contexts. Part 3 considers community-based smoking cessation and tobacco control activities. This volume reveals that no single intervention channel is capable of reaching all smokers. To control smoking, a variety of environmental and social changes need to occur; there needs to be a range of smoking cessation interventions appropriate for many target groups who are in varying stages of readiness to quit smoking. Comprehensive programs must therefore include a variety of strategies available to different types of smokers and deliver on a continuous basis rather than in a single presentation. This volume is the result of a one-day conference in 1991 in Sydney, Australia, entitled, "Interventions for Smokers: An International Perspective." Book contains figures, tables, and a glossary; chapters contain references.

Tollison, Robert D., ed. *Clearing the Air: Perspectives on Environmental Tobacco Smoke.* Lexington, MA: Lexington Books, 1988, 148 pp. ISBN 0-669-18007-6.

This book is about environmental tobacco smoke. The 13 chapters are taken from the proceedings of the 1987 International Conference on Indoor Air Quality in Tokyo, Japan. The conference was hosted by Japan's Council for Environment and Health and sponsored by Philip Morris, Ltd. Topics discussed in the volume include scientific issues regarding exposure to ETS and human health, cigarettes and property rights, a manager's perspective on workplace smoking, ETS and the press, prohibitions and third-party costs, a politician's angle, and politics and meddlesome preferences. At the end of the conference, the organizing committee issued a press release emphasizing the need for further research on the issues of environmental smoke and health efforts. They noted specifically the necessity of giving "priority to solving other public health problems" given the "low probability of proving" the existence of a relationship between ETS and health effects. They concluded that research on various factors in our environment "not just ETS or passive smoking must be promoted." This book presents the basic viewpoint of the tobacco industry. An appendix gives research evidence on environmental smoke. Notes and bibliography are given with each chapter. Information is given about the contributors.

Journal Articles

Air Pollution

General

Anderson, R. C. "Indoors, the Newest Polluted Space." *Pollution Engineering* 24 (April 1, 1992): 58–60.

Ashe, J. T. "Winning the Indoor Air War." *Heating/Piping/Air Conditioning* 64 (September 1992): 71–74.

Brown, Rembert. "Home and Office: Shelter or Threat?" *EPA Journal* 13 (December 1987): 2–5.

Burroughs, H. E. B. "IAQ: An Environmental Factor in the Indoor Habitat." *Heating/Piping/Air Conditioning* 69 (February 1997): 57–60.

Chan, C. C., Leon Vanier, J. W. Martin, and D. T. Williams. "Determination of Organic Contaminants in Residential Indoor Air Using an Adsorption-Thermal Desorption Technique." *Journal of the Air and Waste Management Association* 40 (January 1990): 62–67.

"A Decade of Indoor Air Pollution Problems: University of Cincinnati Engineer Pinpoints the Risks." *Journal of the Air and Waste Management Association* 43 (July 1993): 1012–1013.

Esche, C. A., and J. H. Groff. "ELPAT Program Report: Background and Current Status (October 1996)." *American Industrial Hygiene Association Journal* 58 (January 1997): 59–63.

Hall, S. K., and S. A. Lavite. "Indoor Air Quality of Commercial Buildings." *Pollution Engineering* 20 (June 1988): 54–59.

"IAQ '89 (Indoor Air Quality Conference, San Diego, CA; Products on Display and Abstracts of Papers)" *ASHRAE Journal* 31 (July 1989): 46–60.

"Indoor Air: No Longer Seen as a Safe Haven, Air Indoors Presents Special Pollution Problems." *EPA Journal* 19 (October-December 1993): 6–39.

Jantunen, M. J. "The Indoor Air '93 Congress (Helsinki, Finland, 1993; Selected Papers)." *Atmospheric Environment* (England) 28 (December 1994): 3553–3591.

Kuehn, T. H. "Construction/Renovation Influence on Indoor Air Quality." *ASHRAE Journal* 38 (October 96): 22+.

Levin, H. "Indoor Air Pollution." *ASTM Standardization News* 16 (December 1988): 34–38.

Meckler, M. "Building Renovation and IAQ: A Case Study." *Heating/Piping/Air Conditioning* 66 (July 1994): 69–73.

Nielsen, B. H., and N. O. Breum. "Exposure to Air Contaminants in Chicken Hatching." *American Industrial Hygiene Association Journal* 56 (August 1995): 804–808.

Redpath, D. J. "Hazardous Materials in the Workplace: Things You Never Thought Of." *Energy Engineering* 90, no. 1 (1993): 54–61.

Reese, J. A."ASHRAE Standard 62R: How It Could Affect Segments of the Air Conditioning Industry." *ASHRAE Journal* 39 (January 1997): 30–36.

"Revisions to Standard 69 to Help Improve IAQ (Indoor Air Quality Conference, Baltimore, MD. Oct. 6–8, 1996)" *ASHRAE Journal* 38 (November 1996): 6+.

Rieder, W., and F. Dalfanian. "Simulation of Airborne Pollutant Levels and Infiltration Flows in an Enclosure." *Journal of Solar Energy Engineering* 113 (November 1991): 236–243.

Robinson, John P., and Tibbett L. Speer. "The Air We Breathe." *American Demographics* 17 (June 1995): 24–27+.

Sack, Thomas H., and David H. Steele. *Indoor Air Pollutants from Household Product Sources.* EPA600/4-91/025. Las Vegas, NV: Environmental Monitoring Systems Laboratory, Office of Research and Development, U.S. Environmental Protection Agency, 1991, various paging, microfiche.

Stewart, S. M. "Reaching Agreements on Indoor Air Quality." *ASHRAE Journal* 34 (August 1992): 28–32.

Turner, W. A., and D. Bearg. "Identifying and Avoiding Indoor Air Quality Problems." *Heating/Piping/Air Conditioning* 59 (February 1987): 45–49.

Wendl, D. "Strategic Direction for the IAQ Industry." *Energy Engineering* 93, no. 6 (1996): 64–77.

"What Price Safety?" *Chemistry and Industry,* no. 15 (August 1, 1994): 586.

Wheeler, A. E. "Energy Conservation and Acceptable Indoor Air Quality in the Classroom." *ASHRAE Journal* 34 (April 1992): 26–28+.

Air Quality

Alpers, R., and J. Zaragosa. "Air Quality Sensors for Demand Controlled Ventilation." *Heating/Piping/Air Conditioning* 66 (July 1994): 89–91.

Beckwith, W. "Case History: The Retrofit of a Run-Around Loop for Economical Humidity Control in a Surgery Center." *Energy Engineering* 91, no. 1 (1994): 60–72.

Clark, R. "Indoor Air Quality: Is Dissemination of Information an Adequate Substitution for Regulation?" *ASTM Standard News* 18 (November 1990): 44–50.

Coad, W. J. "Indoor Air Quality: A Designer Parameter." *ASHRAE Journal* 38 (June 1996): 39–40+. Discussion, 38 (September 1996): 16–17.

Collett, C. W., J. A. Ross, and E. M. Sterling. "Quality Assurance Strategies for Investigating IAQ Quality Problems." *ASHRAE Journal* 36 (June 1994): 42–44+.

Curl, S. C., E. L. Joyner, and V. K. Handy. "A Building Owner's Perspective on Indoor Air Quality." *Heating/Piping/Air Conditioning* 65 (August 1993): 61–64+.

Dozier, J. "IAQ and the Classroom." *Heating/Piping/Air Conditioning* 64 (August 1992): 59–62.

Farant, J. P., et al. "Measurement and Impact of Outdoor Air Supplied to Individual Office Building Occupants on Indoor Air Quality." *American Industrial Hygiene Association Journal* 52 (September 1991): 387–392.

Gates, S. D. "Integrating Indoor Air Quality with an HVAC Upgrade." *Energy Engineering* 90, no. 3 (1993): 55–74.

Gould, L. "Special Report: Lift Trucks and Indoor Air Quality." *Modern Materials Handling* 49 (February 1994): 48–51.

"Indoor Air Quality." *ASHRAE Journal* 29 (July 1987): 18–38.

Katzel, J. "A Common Sense Approach to Controlling Indoor Air Quality." *Plant Engineering* 45 (April 18, 1991): 32–38.

Kuehn, T. H. "Construction/Renovation Influence on Indoor Air Quality." *ASHRAE Journal* 38 (October 1996): 22+.

Lizardos, E. J. "Designing HVAC Systems for Optimum Indoor Air Quality." *Energy Engineering* 90, no. 4 (1993): 6–29.

Maroni, Marco, Robert Axelrad, and Alessandro Bacaloni. "NATO's Efforts to Set Indoor Air Quality Guidelines and Standards." *American Industrial Hygiene Association Journal* 56 (May 1995): 499–508.

Meckler, M. "Building Renovation and IAQ: A Case Study." *Heating/Piping/Air Conditioning* 66 (July 1994): 69–73.

———. "Demand-Control Ventilation Strategies for Acceptable IAQ." *Heating/Piping/Air Conditioning* 66 (May 1994): 71–74.

———. "How to Avoid IAQ Problems." *Heating/Piping/Air Conditioning* 63 (February 1991): 35–40.

———. "Indoor Air Quality from Commissioning through Building Operations." *ASHRAE Journal* 33 (November 1991): 42–44+.

Miguel, A. H., F. R. DeAquino Neto, J. N. Cardoso, P. D. Vasconcellos, A. S. Pereira, and K. S. G. Marquez. "Characterization of Indoor Air Quality in the Cities of Sao Paulo and Rio de Janeiro." *Environmental Science and Technology* 29 (February 1995): 338–345.

Moseley, C. "Indoor Air Quality Problems: A Proactive Approach for New Renovated Buildings." *Journal of Environmental Health* 53 (November-December 1990): 19–22.

National Research Council. *Indoor Pollutants.* Washington, DC: National Academy Press, 1981.

Nazaroff, W. "Engineering Solutions to Indoor Air Quality Problems (Special Issue)." *Journal of the Air and Waste Management Association* 46 (September 1996): 805–908.

Newman, J. L. "HVAC System Performance and Indoor Air Quality." *Energy Engineering* 88, no. 3 (1991): 61–78.

Patterson, N. R. "Comfort and Indoor Air Quality." *Heating/Piping/Air Conditioning* 63 (January 1991): 107–111.

Petersen, J. E. "Limitations of Ambient Air Quality Standards in Evaluating Indoor Environments." *American Industrial Hygiene Association Journal* 53 (March 1992): 216–220.

"Research and Development on Indoor Air Quality." *Energy Engineering* 85, no. 2 (1988): 53–59.

Robertson, W., and M. Black. "Indoor Air Quality Challenges for the 90's." *Heating/Piping/Air Conditioning* 63 (February 1991): 26–30.

Scofield, C., and N. H. Des Champs. "Low Temperature Air with High IAQ for Dry Climates." *ASHRAE Journal* 37 (January 1995): 34–40.

Sherwood, V. P. "A Primer on Indoor Air Quality." *Plating and Surface Finishing* 81 (December 1994): 48–49.

Sickles, J. E., et al. *A Summary of Indoor Air Quality Research Through 1984: Project Summary.* EPA/600/S 9-87/020. Research Triangle Park, NC: Environmental Protection Agency, Air and Engineering Research Laboratory, 1987, 2 pp. microfiche.

Sliwinski, Ben J., et al. *Indoor Air Quality Management for Operations and Maintenance Personnel.* USACERL Technical Report, P-91/42. Champaign, IL: U.S. Army Corps of Engineers, Construction Engineering Research Laboratory, 1991, 66 pp. microfiche.

Smith, B., and V. Bristow. "Indoor Air Quality and Textiles: An Emerging Issue." *American Dyestuff Reporter* 83 (January 1994): 37–46.

Smith, J. C. "Schools Resolve IAQ/Humidity Problems with Desiccant Preconditioning." *Heating/Piping/Air Conditioning* 68 (April 1996): 73–78.

Tamblyn, B. T., and S. Khandekar. "IAQ: An Operation and Maintenance Perspective." *ASHRAE Journal* 36 (July 1994): 37–42.

Turner, W. A., and D. Bearg. "Identifying and Avoiding Indoor Air Quality Problems." *Heating/Piping/Air Conditioning* 59 (February 1987): 45–49.

U.S. Environmental Protection Agency. *Report to Congress on Indoor Air Quality, Vol. II: Assessment and Control of Indoor Air Pollution.* EPA 400/1-89-0001C. Washington, DC: EPA, 1989.

"Workshop on Indoor Air Quality." *Risk Analysis* 10 (March 1990): 15–91.

Modeling and Assessment

Austin, B. S., Stanley M. Greenfield, Bruce R. Weir, and Gerald E. Anderson. "Modeling the Indoor Environment." *Environmental Science and Technology* 26 (May 1992): 850–858.

Cal, M. P., et al. "Experimental and Modeled Results Describing the Adsorption of Acetone and Benzene onto Activated Carbon Fibers." *Environmental Progress* 13 (February 1994): 26–30.

Chang, J. C. S., and Z. Guo. "Modeling of the Fast Organic Emissions from a Wood-Finishing Product—Floor Wax." *Atmospheric Environment, Part A* 26A, no. 13 (September 1992): 2365–2370.

Cohen, M. A., P. B. Ryan, Yukio Yanagisawa, and S. K. Hammond. "The Validation of a Passive Sampler for Indoor and Outdoor Concentrations of Volatile Organic Compounds." *Journal of the Air and Waste Management Association* 40 (July 1990): 993–997.

Drye, E. E., et al. "Development of Models for Predicting the Distribution of Indoor Nitrogen Dioxide Concentrations." *JAPCA* 39 (September 1989): 1169–1177.

Eskings, I., Z. Grabaric, and B. S. Grabaric. "Monitoring of Pyrocatechol Indoor Air Pollution." *Atmospheric Environment* (England) 29, no. 10 (May 1995): 1165–1170.

Fellin, P., and R. Otson. "Assessment of the Influence of Climatic Factors on Concentration Levels of Volatile Organic Compounds (VOCs) in Canadian Homes." *Atmospheric Ewnvironment* (England) 28, no. 22 (December 1994): 3581–3586.

Fischer, D., and C. G. Uchrin. "Laboratory Simulation of VOC Entry into Residence Basements from Soil Gas." *Environmental Science and Technology* 30 (August 1996): 2598–2603.

Fischer, M. L., et al. "Factors Affecting Indoor Air Concentrations of Volatile Organic Compounds at a Site of Subsurface Gasoline Contamination." *Environmental Science and Technology* 30 (October 1996): 2948–2957.

Fuessle, R. W., E. D. Brill, and J. C. Liebman. "Air Quality Planning: A General Chance-Constraint Model." *Journal of Environmental Engineering* 113 (February 1987): 106–123.

Furtaw, E. J. Jr., M. D. Pandian, D. R. Nelson, and J. V. Behar. "Modeling Indoor Air Concentrations Near Emission Sources in Imperfectly Mixed Rooms." *Journal of the Air and Waste Management Association* 46 (September 1996): 861–868.

Garbeski, K., and R. G. Sextro. "Modeling and Field Evidence of Pressure-Driven Entry of Soil Gas into a House through Permeable Below Grade Walls." *Environmental Science and Technology* 23 (December 1989): 1481–1487.

Georgopoulos, P. G., et al. "Integrated Exposure and Dose Modeling and Analysis System." *Envirionmental Science and Technology* 31 (January 1997): 17–27.

Hawkins, N. C., et al. "Effects of Selected Process Parameters on Emission Rates of Volatile Organic Chemicals from Carpets." *American Industrial Hygiene Association Journal* 53 (May 1992): 275–282.

Jayjock, M. A. "Back Pressure Modeling of Indoor Air Concentrations from Volatilizing Sources." *American Industrial Hygiene Association Journal* 55 (March 1994): 230–235.

Jayjock, M. A., D. R. Doshi, E. Nungessec, and W. D. Shade. "Development and Evaluation of a Source/Sink Model of Indoor Air Concentrations from Isothiazolone-Treated Wood Used Indoors." *American Industrial Hygiene Association Journal* 56 (June 1995): 546–557.

Jayjock, M. A., and N. C. Hawkins. "A Proposal for Improving the Role of Exposure Modeling in Risk Assessment." *American Industrial Hygiene Association Journal* 54 (December 1993): 733–741.

Kostiainen, R. "Volatile Organic Compounds in the Indoor Air of Normal and Sick Houses." *Atmospheric Environment* (England) 29, no. 6 (April 1995): 693–702.

Lindstrom, A. B., and J. D. Pleil. "A Methodological Approach for Exposure Assessment Studies in Residences using Volatile Organic Compound–Contaminated Water." *Journal of the Air and Waste Management Association* 46 (November 1996): 1058–1066.

Little, J. C., Alfred T. Hodgson, and Ashok J. Gadqil. "Modeling Emissions of Volatile Organic Compounds from New Carpets." *Atmospheric Environment* (England) 28, no. 2 (January 1994): 227–234.

Nazaroff, W. W., and G. R. Cass. "Mathematical Modeling of Indoor Aerosol Dynamics." *Environmental Science and Technology* 23 (February 1989): 157–166.

Nicas, M. "Modeling Respirator Penetration Values with the Beta Distribution: An Application to Occupational Tuberculosis Transmission." *American Industrial Hygiene Association Journal* 55 (June 1994): 515–524.

Park, C., and R. P. Garrison. "Multicellular Model for Contaminant Dispersion and Ventilation Effectiveness with Application for Oxygen Deficiency in a Confined Space." *American Industrial Hygiene Association Journal* 51 (February 1990): 70–78.

Rasouli, F., and T. A. Williams. "Application of Dispersion Modeling to Indoor Gas Release Scenarios." *Journal of the Air and Waste Management Association* 45 (March 1995): 191–195.

Seifert, B., and D. Ullrich. "Methodologies for Evaluating Sources of Volatile Organic Chemicals (VOC) in Homes." *Atmospheric Environment* 21, no. 2 (1987): 395–404.

Sinclair, J. D., L.A. Psota-Kelty, C. J. Weschler, and H. C. Shields. "Measurement and Modeling of Airborne Concentrations and Indoor Surface Accumulation Rates of Ionic Substances at Neenah, Wisconsin." *Atmospheric Environment, Part A* 24A, no. 3 (1990): 627–638.

Stogner, Ronald E., et al. *Two Indoor Air Exposure Modeling Studies, CONTAM Modeling Results and Serial Correlation Effects.* EPA/600/53-91/013. Research Triangle Park, NC: U.S. Environmental Protection Agency, Atmospheric Research and Exposure Assessment Laboratory, 1991, 3 pp. microfiche.

White, W. C., and C. C. Brillhart. "On-site Assessment of Microorganisms and Volatile Organic Compounds." *Energy Engineering* 90, no. 5 (1993): 62–80.

Zhang, Junfeng, Qingci He, and Paul J. Lloy. "Characteristics of Aldehydes: Concentrations, Sources, and Exposures for Indoor and Outdoor Residential Microenvironments." *Environmental Science and Technology* 28 (January 1994): 146–152.

Zhang, Junfeng, William E. Wilson, and Paul J. Lloy. "Indoor Air Chemistry: Formation of Organic Acids and Aldehydes." *Environmental Science and Technology* 28 (October 1994): 1975–1982.

Laws and Regulations

Diamond, Mark. "Liability in the Air: The Threat of Indoor Pollution." *ABA (American Bar Association) Journal* 73 (November 1, 1987): 78–80+.

Eisenstein, H. "IAQ: Who is Legally Responsible?" *Heating/Piping/Air Conditioning* 64 (August 1992): 43–46+.

Hodgson, M. J., and C. A. Hess. "Doctors, Lawyers, and Building-Associated Diseases." *ASHRAE Journal* 34 (February 1992): 25–32.

Miro, C. R., and J. E. Cox. "Federal Regulations Proposed for Indoor Air in Nation's Workplaces." *ASHRAE Journal* 36 (June 1994): 16.

Paskai, S. S. "Problems in Regulating Office Air Quality: A Critique of One State's Attempt." *American Industrial Hygiene Association Journal* 55 (March 1994): 255–256.

U.S. Congress. House Committee on Government Operations. Environment, Energy, and Natural Resources Subcommittee. *Potential Health Risks from Carpets and Carpeting Material: Hearing, June 11, 1993.* 103d Cong., 1st sess. Washington, DC: GPO, 1993, 244 pp.

U.S. Congress. House Committee on Public Works and Transportation. Subcommittee on Aviation. *Airliner Cabin Air Quality: Hearing, May 18, 1994.* 103d Cong., 2d sess. Washington, DC: GPO, 1994, 516 pp.

U.S. Congress. House Committee on Science, Space, and Technology. Subcommittee on Environment. *H. R. 1066, the Indoor Air Quality Act of 1991: Hearing, May 9, 1991.* 102d Cong., 1st sess. Washington, DC: GPO, 1991, 395 pp.

U.S. Congress. House Committee on Science, Space, and Technology. Subcommittee on Natural Resources, Agriculture Research, and Environment. *H. R. 1530—the Indoor Air Quality Act of 1989: Hearings, July 20 and September 27, 1989.* 101st Cong., 1st sess. Washington, DC: GPO, 1990, 498 pp.

———. *Indoor Air Quality Act of 1988: Hearing, September 28, 1988.* 100th Cong., 2d sess. Washington, DC: GPO, 1989, 266 pp.

U.S. Congress. Senate Committee on Environment and Public Works. *Indoor Air Pollution: Hearing, August 5, 1985, on S. 1198, a Bill to Establish in the Environmental Protection Agency a Program of Research on Indoor Air Quality, and for Other Purposes.* 99th Cong., 1st sess. Washington, DC: GPO, 1985, 68 pp.

U.S. Congress. Senate Committee on Environment and Public Works. Subcommittee on Clean Air and Nuclear Regulation. *Pending Indoor Air Quality and Radon Abatement Legislation: Hearing, May 25, 1993, on S. 656, a Bill to Provide for Indoor Air Pollution Abatement, Including Indoor Radon Abatement and for Other Purposes, and S. 657.* 103d Cong., 1st sess. Washington, DC: GPO, 1993, 306 pp.

U.S. Congress. Senate Committee on Environment and Public Works. Subcommittee on Superfund, Ocean, and Water Protection. *Indoor Air Quality Act of 1989: Hearing, May 3, 1989, on S. 657, a Bill to Authorize a National Program to Reduce the Threat to Human Health Posed by Exposure to Contaminants in the Air Indoors.* 101st Cong., 1st sess. Washington, DC: 1989, 291 pp.

U.S. Congress. Senate Committee on Governmental Affairs. Ad Hoc Subcommittee on Consumer and Environmental Affairs. *New Research on the Potential Health Risks of Carpets: Hearing, October 1, 1992.* 102d. Cong., 2d sess. Washington, DC: GPO, 1994, 221 pp.

U.S. Environmental Protection Agency. *Managing Indoor Air Quality Risks: Pilot Study on Indoor Air Quality: Report on a Meeting Held in St. Michaels, Maryland, October 25–27, 1989.* Washington, DC: EPA, 1990, 200 pp.

Weymueller, C. R. "OSHA Rules Aim to Clear Air." *Welding Design and Fabrication* 62 (December 1989): 33–36.

Worsnop, Richard L. "Indoor Air Pollution: Would Tougher Regulations Reduce Health Problems?" *CQ Researcher* 5 (October 27, 1995): 945–967.

Health Risks

Adkins, J. E. Jr., and N. W. Henry III. "Industrial Hygiene." *Analytical Chemistry* 67 (June 15, 1995): 349R–376R.

———. "Industrial Hygiene." *Analytical Chemistry* 65 (June 15, 1993): 133R–155R.

Attwood, P., et al. "A Study of the Relationship between Airborne Contaminants and Environmental Factors in Dutch Swine Confinement Buildings." *American Industrial Hygiene Association Journal* 48 (August 1987): 745–751.

Beck, Joe E. "It's Time for a Breath of Fresh Air: Environmental Health Professionals Have Been too Slow to Address Indoor Air Quality Concerns." *Occupational Health and Safety* 64 (February 1996): 25+.

Cain, W. S., Jonathan M. Samet, and Michael Hodgson. "The Quest for Negligible Health Risk from Indoor Air." *ASHRAE Journal* 37 (July 1995): 38–44.

Cuijpers, C. E. J., G. M. H. Swaen, G. Wesseling, F. Sturmans, and E. F. M. Wouters. "Adverse Effects of the Indoor Environment on

Respiratory Health in Primary School Children." *Environmental Research* 68 (January 1995): 11–23.

Davidoff, Linda Lee. "Multiple Chemical Sensitivities (MCS): Also Known as Chemical Hypersensitivity, Total Allergy Syndrome, and Environmental Illness, MCS May Be the Ultimate Twentieth-Century Illness Affecting as Much as 15 Percent of the Population." *Amicus Journal* 11 (Winter 1989): 12–23.

Epstein, E. "Protecting Workers at Composting Facilities." *BioCycle* 37 (September 1996): 69–70+.

Fotos, C. P. "Flight Attendants Question Health Risk of Recirculated Air in Newer Cabins." *Aviation Week and Space Technology* 135 (November 25, 1991): 79+.

Fulwiler, R. D., and R. J. Hackman. "EPA's Intervention into Workplace Health and Safety—the Other OSHA." *American Industrial Hygiene Association Journal* 51 (July 1990): A490–A495.

Fumento, Michael. "Sick of It All: People with 'Multiple Chemical Sensitivity' Are Definitely Suffering: The Question Is, Why?" *Reason* 28 (June 1996): 20–26.

Gabbay, Jacob, Orna Bergerson, Nissim Levi, Shmuel Brenner, and Ilana Eli. "Effect of Ionization on Microbial Air Pollution in the Dental Clinic." *Environmental Research* 52 (June 1990): 99–106.

George, D. K., M. R. Flynn, and R. L. Harris. "Autocorrelation of Interday Exposure at an Automobile Assembly Plant." *American Industrial Hygiene Association Journal* 56 (December 1995): 1187–1194.

Gill, K. E., and A. L. Wozniak. "Hospital Gets IAQ Checkup." *Heating/Piping/Air Conditioning* 65 (August 1993): 43–47+.

Giroux, Denise, Gilles Lapointe, and Marc Baril. "Toxiciological Index and the Presence in the Workplace of Chemical Hazards for Workers Who Breast-Feed Infants." *American Industrial Hyygiene Association Journal* 53 (July 1992): 471–474.

Grant, Yola. "Indoor Air Pollution: An Emergent Occupational Hazard." *Canadian Journal of Women and the Law* 4, no. 1 (1990): 235–251.

Greenberg, Michael R. "Indoor Air Quality: Protecting Public Health through Design, Planning, and Research." *Journal of Architectural and Planning Research* 3 (August 1986): 253–261.

Greene, R. E., and P. L. Williams. "Indoor Air Quality Investigation Protocols." *Journal of Environmental Health* 59 (October 1996): 6–13+.

Hayes, S. R. "Estimating the Effects of Being Indoors on Total Personal Exposure to Outdoor Air Pollution." *JAPCA* 39 (November 1989): 1453–1461.

Heida, Henk, Frans Bartman, and Saskia C. van der Zee. "Occupational Exposure and Indoor Air Quality Monitoring in a Composting Facility." *American Industrial Hygiene Association Journal* 56 (January 1995): 39–43.

Hileman, B. "Multiple Chemical Sensitivity." *Chemical and Engineering News* 69 (July 22, 1991): 26–42. Discussion, 69 (September 9, 1991): 3+; 69 (September 23, 1991); 3, 69 (October 21, 1991): 3.

Hodgson, M. A. "Health Risks of Indoor Pollutants." In *Engineering Solutions to Indoor Air Problems.* Atlanta: ASHRAE, 1988, pp. 284–293.

Indoor Air-Assessment: Indoor Concentrations of Environmental Carcinogens. EPA/600/8-90/042. Research Triangle Park, NC: Environmental Criteria and Assessment Office, Office of Health and Environmental Assessment, Office of Research and Development, U.S. Environmental Protection Agency, 1991, 42 pp. microfiche.

"Indoor Air: Dirty, Contaminated, and Yours to Breathe." *Journal of American Insurance* 65 (First Quarter 1989): 16–20.

Johnson, P. W. "Avoiding Headaches from In-Plant Air Problems." *Pollution Engineering* 27 (March 1995): 30–33.

Laird, Russell. "Sick of the System?" *Occupational Health and Safety* 63 (September 1994): 65–66+.

Landrigan, P. J., D. G. Graham, and R. D. Thomas. "Strategies for the Prevention of Environmental Neurotoxic Illness." *Environmental Research* 61 (April 1993): 157–163.

Mann, J. H. Jr. "An Industrial Hygiene Evaluation of Beet Sugar Processing Plants." *American Industrial Hygiene Association Journal* 51 (June 1990): 313–318.

Martin, J. R. "The Sensitive Individual and the Indoor Environment: Case Study." *American Industrial Hygiene Association Journal* 56 (November 1995): 1121–1126.

Moschandreas, D. J., and P. E. Chang. "On the Use of a Risk Ladder: Linking Public Perception of Risks Associated with Indoor

Air with Cognitive Elements and Attitudes Toward Risk Reduction." *Atmospheric Environment* (England) 28, no. 19 (November 1994): 3093–3098.

Muller, W. J., and V. H. Schaeffer. "A Strategy for the Evaluation of Sensory and Pulmonary Irritation Due to Chemical Emissions from Indoor Sources." *Journal of the Air and Waste Management Association* 46 (September 1996): 808–812.

Muttamara, S., and K. U. Alwis. "Health Impacts of Garage Workers: A Preliminary Study." *Journal of Environmental Health* 56 (May 1994): 19–24.

New Jersey Department of Health. Division of Epidemiology, Environmental, and Occupational Health Services. *Chromium Medical Surveillance Project: Summary of Final Technical Report.* Trenton, NJ: 1994, 8+ pp.

Nielsen, B. H., and N. O. Breum. "Exposure to Air Contaminants in Chicken Catching." *American Industrial Hygiene Association Journal* 56 (August 1995): 804–808.

Read, R., and C. Read. "Breathing Can Be Hazardous to Your Health." *New Scientist* 129 (February 23, 1991): 34–37.

Ricci, P. F. "Mortality, Air Pollution, and Energy Production: Uncertainty and Casualty." *Journal of Energy Engineering* 116 (December 1996): 148–162.

Schwartz, J., C. Spix, H. E. Wichmann, and E. Malin. "Air Pollution and Acute Respiratory Illness in Five German Communities." *Environmental Research* 56 (October 1991): 1–14.

Tancrede, M., R. Wilson, L. Zeise, and E. A. C. Crouch. "The Carcinogenic Risk of Some Organic Vapors Indoors: A Theoretical Survey." *Atmospheric Environment* 21, no. 10 (1987): 2187–2205.

Turk, A. R., and E. M. Poulakos. "Practical Approaches for Healthcare: Indoor Air Quality Management." *Energy Engineering* 93, no. 5 (1996): 12–79.

U.S. Congress. House Committee on Energy and Commerce. Subcommittee on Health and the Environment. *Indoor Air Pollution: Hearing, April 10, 1991.* 102d Cong., 1st sess. Washington, DC: GPO, 1991, 291 pp.

———. *Indoor Air Pollution: Hearing November 1, 1993, on H. R. 2919, a Bill to Amend the Public Health Service Act to Authorize a National Program to Reduce the Threat to Human Health Posed by Expo-*

sure to Contaminants in the Air Indoors. 103d Cong., 1st sess. Washington, DC: GPO, 1994, 139 pp.

———. *Indoor Air Pollution in Schools: Hearing, March 18, 1993.* 103d Cong., 1st sess. Washington, DC: GPO, 1993, 73 pp.

U.S. Environmental Protection Agency. *Indoor Air Assessment: Methods of Analysis for Environmental Carcinogens.* EPA/600/8-90/041. Research Triangle Park, NC: Environmental Criteria Assessment Office, Office of Health and Environmental Assessment, Office of Research and Development, 1990, 37 pp.

"Variable Frequency Drives Help Hospitals Achieve Air Quality Improvements." *Heating/Piping/Air Conditioning* 68 (December 1996): 9–10+.

Yao, C., D. C. Krueger, K. R. Loos, and J. W. Koehn. "Collection and Determination of 1,3-Butadiene Using Passive Dasimeters and Automatic Thermal Desorption." *American Industrial Hygiene Association Journal* 58 (January 1997): 44–50.

Yu, R. C., and R. J. Sherwood. "The Relationship between Urinary Elimination, Airborne Concentration, and Radioactive Hand Contamination for Workers Exposed to Uranium." *American Industrial Hygiene Association Journal* 57 (July 1996): 615–620.

Zummo, S. M., and M. H. Karol. "Indoor Air Pollution: Acute Adverse Health Effects and Host Susceptibility." *Journal of Environmental Health* 58 (January-February 1996): 25–29.

Ambient Air Pollution

Ozone

Bedi, J. F., S. M. Horvath, and D. M. Drechsler-Parks. "Adaptation by Older Individuals Repeatedly Exposed to 0.45 Parts Per Million Ozone for Two Hours." *JAPCA* 39 (February 1989): 194–199.

Boeniger, M. F. "Use of Ozone Generating Devices to Improve Indoor Air Quality." *American Industrial Hygiene Association Journal* 56 (June 1995): 590–598.

Druzik, J. R., et al. "The Measurement and Model Predictions of Indoor Ozone Concentrations in Museums." *Atmospheric Environment, Part A* 24A, no. 7 (1990): 1813–1823.

Fahy, J. V., H. H. Wong, J. T. Liu, and H. A. Boushey. "Analysis of Induced Sputum after Air and Ozone Exposures in Healthy Subjects." *Environmental Research* 70 (August 1995): 77–83.

Hayes, S. R. "Use of an Indoor Air Quality Model (IAQM) to Estimate Indoor Ozone Levels." *Journal of the Air and Waste Management Association* 41 (February 1991): 161–170.

Liu, L-J.S., Melvin P. Olson III, George A. Allen, and Petros Koutrakis. "Evaluation of the Harvard Ozone Passive Sampler on Human Subjects Indoors." *Environmental Science and Technology* 28 (May 1994): 915–923.

Reisenauer, C. S., J. Q. Koenig, M. S. McManus, M. S. Smith, Greg Kusic, and W. E. Pierson. "Pulmonary Response to Ozone Exposure in Healthy Individuals Aged 55 Years or Greater." *JAPCA* 38 (January 1988): 51–55.

Reiss, Richard, P. Barry Ryan, and Petros Koutrakis. "Modeling Ozone Deposition onto Indoor Residential Surfaces." *Environmental Science and Technology* 28 (March 1994): 504–513.

Reiss, Richard, P. B. Ryan, P. Koutrakis, and S. J. Tibbetts. "Ozone Reactive Chemistry on Interior Latex Paint." *Environmental Science and Technology* 29 (August 1995): 1906–1912.

Weschler, C. J., Michael Brauer, and Petros Koutrakis. "Indoor Ozone and Nitrogen Dioxide: A Potential Pathway to the Generation of Nitrate Radicals, Dinitrogen Pentaoxide, and Nitric Acid Indoors." *Environmental Science and Technology* 26 (January 1992): 179–184.

Weschler, C. J., A. T. Hodgson, and John D. Wooley. "Indoor Chemistry: Ozone, Volatile Organic Compounds, and Carpets." *Environmental Science and Technology* 26 (December 1992): 2371–2377.

Weschler, C. J., H. C. Shields, and D. V. Naik. "Indoor Ozone Exposure." *JAPCA* 39 (December 1989): 1562–1568.

Nitrogen Oxide

Brauer, Michael, P. B. Ryan, H. H. Suh, Petros Koutrakis, and John D. Spangler. "Measurements of Nitrous Acid inside Two Research Houses." *Environmental Science and Technology* 24 (October 1990): 1521–1527.

Brunekreef, Bert, Danny Houthuijs, Lyanne Dijkstra, and Jan S. M. Boleij. "Indoor Nitrogen Dioxide Exposure and Children's Pulmonary Function." *Journal of the Air and Waste Management Association* 40 (September 1990): 1252–1256.

Culotta, E., and D. E. Kashland Jr. "NO News Is Good News" *Science* 258 (December 18, 1992): 1862–1865.

Febo, A., and C. Perrino. "Measurement of High Concentration of Nitrous Acid Inside Automobiles." *Atmospheric Environment* (England) 29, no. 3 (February 1995): 345–351.

———. "Prediction and Experimental Evidence for High Air Concentration of Nitrous Acid in Indoor Environments." *Atmospheric Environment, Part A,* 25A, nos. 5–6 (1991): 1055–1061.

Feldman, P. L., et al. "The Surprising Life of Nitric Oxide (Special Report)." *Chemical and Engineering News* 71 (December 20, 1993): 26–38. Discussion, 72 (March 14, 1994): 4.

Goldstein, I. F., L. R. Andrews, and Diana Hartel. "Assessment of Human Exposure to Nitrogen Dioxide, Carbon Monoxide, and Respirable Particulates in New York Inner-City Residences." *Atmospheric Environment* 22, no. 10 (1988): 2127–2139.

Leaderer, B. P., R. T. Zagraniski, Marianne Beswick, and Jan A. J. Stolwijk. "Predicting NO_2 Levels in Residences Based upon Sources and Source Use: A Multivariate Model." *Atmospheric Environment* 21, no. 2 (1987): 361–368.

Lee, K., Y. Yanagisawa, and J. D. Spengler. "Carbon Monoxide and Nitrogen Dioxide Levels in an Indoor Skating Rink with Mitigation Methods." *Journal of the Air and Waste Management Association* 43 (May 1993): 769–771.

Melia, R. J. W., S. Chinn, and R. J. Rona. "Indoor Levels of NO_2 Associated with Gas Cookers and Kerosene Heaters in Inner City Areas of England." *Atmospheric Environment, Part B* 24B, no. 1 (1990): 177–180.

Moschandreas, D. J., S. M. Relwani, and E. H. Luebcke. "Fugitive Emissions of NO_2 from Vented Gas Appliances in Residences—A Pilot Study." *Journal of the Air and Waste Management Association* 40 (March 1990): 359–361.

Moschandreas, D. J., S. M. Relwani, K. C. Taylor, and J. D. Mulik. "A Laboratory Evaluation of a Nitrogen Dioxide Personal Sampling Device." *Atmospheric Environment, Part A* 24A, no. 11 (1990): 2807–2811.

Noy, Dook, Bert Brunekreef, J. S. M. Boleij, Danny Houthuijs, and Rian De Koning. "The Assessment of Personal Exposure to Nitrogen Dioxide in Epidemiological Studies." *Atmospheric Environment, Part A* 24A, no. 12 (1990): 2903–2909.

Petreas, Myrto, Kai-Shen Liu, Bei-Hung Chang, Steven B. Hayward, and Ken Sexton. "A Survey of Nitrogen Dioxide Levels Measured Inside Mobile Homes." *JAPCA* 38 (May 1988): 647–651.

Pitts, J. N. Jr., H. W. Biermann, E. C. Tuazon, M. Green, W. D. Long, and A. M. Winer. "Time-Resolved Identification and Measurement of Indoor Air Pollutants by Spectroscopic Techniques: Gaseous Nitrous Acid, Methanol, Formaldehyde, and Formic Acid." *JAPCA* 39 (October 1989): 1344–1347.

Ryan, P. B., M. L. Soczek, R. D. Treitman, J. D. Spangler, and I. W. Billick. "The Boston Residential NO_2 Characterization Study: Survey Methodology and Population Concentration Estimates." *Atmospheric Environment* 22, no. 10 (1988): 2115–2125.

Salmon, L. G., W. W. Nazaroff, M. P. Ligocki, M. C. Jones, and G. R. Cass. "Nitric Acid Concentrations in Southern California Museums." *Environmental Science and Technology* 24 (July 1990): 1004–1113.

Smil, Vaclav. "Global Population and the Nitrogen Cycle." *Scientific American* 277 (July 1997): 76–81.

Spicer, C. W., D. V. Kenny, G. F. Ward, I. H. Billick, and N. P. Leslie. "Evaluation of NO_2 Measurement Methods for Indoor Air Quality Applications." *Journal of the Air and Waste Management Association* 44 (February 1994): 163–168.

U.S. Environmental Protection Agency. *Air Quality Criteria for Oxides of Nitrogen.* EPA-600/8-82-026F. Research Triangle Park, NC: EPA, 1982.

Young, S. "The Body's Vital Poison." *New Scientist* 137 March 13, 1993): 36–40.

Carbon Monoxide

Alm, S., A. Reponen, K. Mukala, P. Pasanen, J. Tuomisto, and M. J. Jantunen. "Personal Exposures of Preschool Children to Carbon Monoxide: Roles of Ambient Air Quality and Gas Stoves." *Atmospheric Environment* (England) 28, no. 22 (December 1994): 3577–3580.

Barinaga, M. "Carbon Monoxide: Killer to Brain Messenger in One Step." *Science* 259 (January 15, 1993): 309.

Boudreau, D. R., M. P. Spadafora, L. R. Wolf, and E. Siegel. "Carbon Monoxide Levels during Indoor Sporting Events—Cincinnati, 1992–1993." *Journal of Environmental Health* 57 (November 1994): 21–22.

Fernandez-Bremauntz, A. A., and M. R. Ashmore. "Exposure of Commuters to Carbon Monoxide in Mexico City—I. Measurement of In-Vehicle Concentrations." *Atmospheric Environment* (England) 29, no. 4 (March 1995): 525–532.

Garsik, D. A. "Evaluation of Airborne Carbon Monoxide Exposure Monitoring Program in Produce Cooler Operations (Palm Beach County, Florida)." *Journal of Environmental Health* 57 (May 1995): 19–22.

He, Fengsheng, et al. "Evaluation of Brain Function in Acute Carbon Monoxide Poisoning with Multimodality Evoked Potentials." *Environmental Research* 60 (February 1993): 213–226.

Koushki, P. A., K. H. Al-Dhowalia, and S. A. Niaizi. "Vehicle Occupant Exposure to Carbon Monoxide." *Journal of the Air and Waste Management Association* 42 (December 1992): 1603–1608.

Nagda, N. L., M. D. Koontz, and A. G. Konheim. "Carbon Dioxide Levels in Commercial Airliner Cabins." *ASHRAE Journal* 33 (August 1991): 35–38.

Schwab, M. "An Examination of the Intra-SMSA Distribution of Carbon Monoxide Exposure." *Journal of the Air and Waste Management Association* 40 (March 1990): 331–336.

Stern, C. H., D. R. Jaasma, J. W. Shelton, and Gary Satterfield. "Parametric Study of Fireplace Particulate Matter and Carbon Monoxide Emissions." *Journal of the Air and Waste Management Association* 42 (June 1992): 777–783.

"Unintentional Carbon Monoxide Poisonings in Residential Settings." *Journal of Environmental Health* 58 (January-February 1996): 36–37.

Verma, Ajay, David J. Hersch, Charles E. Glatt, G. V. Ronnett, and S. H. Snyder. "Carbon Monoxide: A Putative Neural Messenger." *Science* 259 (January 15, 1993): 381–384.

Wallace, Lance, Jacob Thomas, David Mage, and Wayne Ott. "Comparison of Breath CO, CO Exposure, and Coburn Model Predictions in the U.S. EPA Washington-Denver (CO) Study." *Atmospheric Environment* 22, no. 10 (1988): 2183–2193.

Carbon Dioxide

Batterman, S., and C. Peng. "TVOC and CO_2 Concentrations as Indicators in Indoor Air Quality Studies." *American Industrial Hygiene Association Journal* 56 (January 1995): 55–65.

Burke, M. "Are Oxyfuels Good for Us?" *New Scientist* (July 15, 1995): 24–27.

Jacobs, D. E., and M. S. Smith. "Exposures to Carbon Dioxide in the Poultry Processing Industry." *American Industrial Hygiene Association Journal* 49 (December 1988): 624–629.

Jankovic, J. T., Robert Ihle, and D. O. Vick. "Occupant Generated Carbon Dioxide as a Measure of Dilution Ventilation Efficiency." *American Industrial Hygiene Association Journal* 57 (August 1996): 756–759.

Norberg-Bohm, V. "From the Inside Out (Reducing CO_2 Emissions in the Buildings Sector). *Environment* 33 (April 1991): 16–20+.

Stonier, R. T. "CO_2: Powerful IAQ Diagnostic Tool." *Heating/Piping/Air Conditioning* 67 (March 1995): 88–90+.

Tiwari, P., and J. Parikh. "Cost of CO_2 Reduction in Building Construction." *Energy* 20 (June 1995): 531–547.

Sulfur Dioxide

Hisham, M. W. M., and D. Grosjean. "Sulfur Dioxide, Hydrogen Sulfide, Total Reduced Sulfur, Chlorinated Hydrocarbons, and Photochemical Oxidants in Southern California Museums." *Atmospheric Environment, Part A* 25A, no. 8 (1991): 1497–1505.

Tobacco Smoke

Ashley, D. L., M. A. Bonin, Brent Hamar, and M. A. McGeehin. "Removing the Smoking Confounder from Blood Volatile Organic Compounds Measurements." *Environmental Research* 71 (October 1995): 39–45.

Aviado, D. M. "Cardiovascular Disease and Occupational Exposure to Environmental Tobacco Smoke." *American Industrial Hygiene Association Journal* 57 (March 1996): 285–294; Discussion, 57 (November 1996): 1076–1077.

Ballard, J. A. "Health Effects of Passive Smoking." *Professional Safety* 37 (October 1992): 28–32.

Barinaga, M. "[California Governor Pete] Wilson Slashes Spending for Anti-Smoking Effort." *Science* 255 (March 13, 1992): 1348–1349.

Barlow, Stephanie. "Up in Smoke: No Ifs, Ands, or Butts—Smoking Is on Its Way Out." *Entrepreneur* 22 (August 1994): 126+.

Barnes, Deborah E. "Industry-Funded Research and Conflict of Interest: An Analysis of Research Sponsored by the Tobacco Industry through the Center for Indoor Air Research." *Journal of Health Politics, Policy, and Law* 21 (Fall 1996): 515–542.

Bowers, Mollie H. "What Labor and Management Need to Know about Workplace Smoking Cases." *Labor Law Journal* 43 (January 1992): 40–49.

Brown, Barbara A. "Work Place Smoking Restrictions: An Update." *Employment Relations Today* 14 (Spring 1987): 31–36.

Burkhard, Michael S., and M. Allison Despard. "Cigarette Classification a Burning Issue: New Evidence Could Lead to 'Drug' Classification for Cigarettes." *Loyola Consumer Law Reporter* 6 (Summer 1994): 116–121.

Cain, W. S., Tarik Tosun, Lai-Chu See, and Brian Leaderer. "Environmental Tobacco Smoke: Sensory Reactions of Occupants." *Atmospheric Environment* 21, no. 2 (1987): 347–353.

Caka, F. M., et al. "An Intercomparison of Sampling Techniques for Nicotine in Indoor Environment." *Environmental Science and Technology* 24 (August 1990): 1196–1203.

Chaloupka, Frank J., and Henry Saffer. "Clean Indoor Air Laws and the Demand for Cigarettes." *Contemporary Policy Issues* 10 (April 1992): 72–83.

Cloud, David S. "Tobacco Industry Losing Allies as Congress Eyes Health Tax: With the Public Strongly Supporting Higher Levies on Cigarettes, the Real Battle May Be Over How High They Should Be." *Congressional Quarterly Weekly Report* 52 (April 23, 1994): 985–989.

Cohen, J. "Tobacco Money Lights Up a Debate." *Science* 272 (April 26, 1996): 488–494; Discussion, 272 (May 31, 1996): 1247.

Cooper, Mary H. "Regulating Tobacco: Can the FDA Break America's Smoking Habit?" *CQ Researcher* 4 (September 30, 1994): 841–863.

Eatough, D. J., F. M. Caka, John Crawford, Scott Braithwaite, L. D. Hanson, and E. A. Lewis. "Environmental Tobacco Smoke in

Commercial Aircraft." *Atmospheric Environment, Part A* 26A, no. 12 (August 1992): 2211–2218.

Fox, John C., and Bernadette M. Davison. "Smoking in the Workplace: Accommodating Diversity." *California Western Law Review* 25, no. 2 (1988–1989): 215–237.

Fruchtman, D. J. "Smoking in Restaurants—Chasing the Air." *Heating/Piping/Air Conditioning* 64 (July 1992): 75–76. Discussion, 64 (December 1992): 32.

Gallup, George Jr., and Frank Newport. "Many Americans Favor Restrictions on Smoking in Public Places." *Gallup Poll Monthly* (July 1990): 19–27.

Hanley, N.J., S. G. Colosimo, C. M. Axelrad, Randall Harris, and D. W. Sepkovic. "Biochemical Validation of Self-Reported Exposure to Environmental Tobacco Smoke." *Environmental Research* 49 (June 1989): 127–135.

Heavner, D. L., M. W. Ogden, and Paul R. Nelson. "Multisorbent Thermal Desorption/Gas Chromatography/Mass Selective Detection Method for the Determination of Target Volatile Organic Compounds in Indoor Air." *Environmental Science and Technology* 26 (September 1992): 1737–1746.

Henschen, M., et al. "The Internal Dose of Passive Smoking at Home Depends on the Size of the Dwelling." *Environmental Research* 72 (January 1997): 65–71.

Hu, Teh-Wei, Jushan Bai, Theodore Keeler, Paul Barnett, and Hai-Yen Sung. "The Impact of California Proposition 99, a Major Anti-Smoking Law, on Cigarette Consumption." *Journal of Public Health Policy* 15 (Spring 1994): 26–36.

Huber, Gary L., Robert E. Brockie, and Vijay K. Mahajan. "Smoke and Mirrors: The EPA's Flawed Study of Environmental Tobacco Smoke and Lung Cancer." *Regulation* (Cato Institute) 16, no. 3 (1993): 44–54.

"Indoor Air: No Longer Seen as a Safe Haven, Air Indoor Presents Special Pollution Problems." *EPA Journal* 19 (October-December 1993): 6–39.

Jacobson, Peter D., et al. "The Politics of Antismoking Legislation." *Journal of Health Politics, Policy, and Law* 18 (Winter 1993): 787–819.

Kaiser, J. "Another Look at Secondhand Smoke." *Science* 270 (November 10, 1995): 903.

Kamens, Richard, Chang-Te Lee, Russell Wiener, and David Leith. "A Study to Characterize Indoor Particles in Three Non-Smoking Homes." *Atmospheric Environment, Part A* 25A, nos. 5–6 (1991): 939–948.

Kemper, Vicki. "The Inhalers: They May Not Smoke Tobacco Products, But Some in Congress Are Addicted to the Industry's Money." *Common Cause Magazine* 21 (Spring 1995): 18–23.

Klepeis, N. E., Wayne R. Ott, and Paul Switzer. "A Multiple-Smoker Model for Predicting Indoor Air Quality in Public Lounges." *Environmental Science and Technology* 30 (September 1996): 2813–2820.

Koutrakis, Petros, Susan L. K. Briggs, and Brian P. Leaderer. "Source Apportionment of Indoor Aerosols in Suffolk and Onondaga Counties, New York." *Environmental Science and Technology* 26 (March 1992): 521–527.

Kumagai, J. "Exposing the Dangers of Tobacco Smoke." *Physics Today* 48 (October 1995): 59–60.

Landsberger, S., et al. "Determination of Airborne Calcium in Environmental Tobacco Smoke by Instrumental Neutron Activation Analysis with a Compton Suppression System." *Analytical Chemistry* 65 (June 1, 1993): 1506–1509.

Leaderer, B. P., and S. K. Hammond. "Evaluation of Vapor-Phase Nicotine and Respirable Suspended Particle Mass as Markers for Environmental Tobacco Smoke." *Environmental Science and Technology* 25 (April 1991): 770–777.

Lesmes, George, Jeffrey Lakier, and Alan Mintz. "Getting Employees to Say No to Smoking." *Business and Health* 11 (March 1993): 42–44+.

Li, C.-S., Wen-Hai Lin, and Fu-Tien Jenq. "Characterization of Outdoor Submicron Particles and Selected Combustion Sources of Indoor Particles." *Atmospheric Environment, Part B* 27B (December 1993): 413–424.

Lippert, Julie E. "Prenatal Injuries from Passive Tobacco Smoke: Establishing a Cause of Action for Negligence." *Kentucky Law Journal* 78, no. 4 (1989–1990): 865–878.

Mahanama, K. R. R., and J. M. Daisey. "Volatile N-Nitrosamines in Environmental Tobacco Smoke: Sampling, Analysis, Emission Factors, and Indoor Air Exposures." *Environmental Science and Technology* 30 (May 1996): 1477–1484.

Mallory, Maria. "Is the Smoking Lamp Going out for Good? Never Has the Tobacco Industry Been So Beleaguered Nor Its Foes So Bold." *Business Week* (April 11, 1994): 30–31.

Martonen, T. B. "Deposition Patterns of Cigarette Smoke in Human Airways." *American Industrial Hygiene Association Journal* 53 (January 1992): 6–18.

Massey, Marilyn M., Gayle Boyd, Margaret Mattson, and Marcia Feinleib. "Inventory Surveys on Smoking." *Public Health Reports* 102 (July-August 1987): 430–438.

Miesner, E. A., S. N. Rudnick, F. Hu, J. D. Spengler, L. Preller, H. Ozkaynak, and W. Nelson. "Particulate and Nicotine Sampling in Public Facilities and Offices." *JAPCA* 39 (December 1989): 1577–1582.

Miller, A. "Pulmonary Function in Asbestos and Asbestos-Related Pleural Disease." *Environmental Research* 61 (April 1993): 1–18.

Moschandreas, D. J., and S. M. Relwani. "Perception of Environmental Tobacco Smoke Odors: An Olfactory and Visual Response." *Atmospheric Environment, Part B* 26B (September 1992): 263–269.

Mullooly, John P., Katharina Schuman, Victor J. Stevens, Russell Glasgow, and Thomas Vogt. "Smoking Behavior and Attitudes of Employees of a Large HMO before and after a Work Site Ban on Cigarette Smoking." *Public Health Reports* 105 (November-December 1990): 623–628.

Mundell, I. "Peering through the Smoke Screen (Cigarette Smoking Appears to Protect against a Number of Diseases)." *New Scientist* 140 (October 9, 1993): 14–15.

Nagda, N. L., M. D. Koontz, and A. G. Konheim. "Measurement of Cabin Air Quality Aboard Commercial Airliners." *Atmospheric Environment, Part A* 26A, no. 12 (August 1992): 2203–2210.

Nelson, P. R., D. L. Heavner, B. B. Collie, K. C. Malolo, and M. W. Ogden. "Effect of Ventilation and Sampling Time on Environmental Tobacco Smoke Component Ratios." *Environmental Science and Technology* 26 (October 1992): 1909–1921.

Nixon, Judy C., and Judy F. West. "The Ethics of Smoking Policies." *Journal of Business Ethics* 8 (June 1989): 409–414.

"No Smoking: The Social Revolution Sweeping America." *Business Week* (July 27, 1987): 40–43+.

Ogden, M. W., and K. C. Maiolo. "Collection and Determination of Salanesol as a Tracer of Environmental Tobacco Smoke in Indoor Air." *Environmental Science and Technology* 23 (September 1989): 1148–1154.

Oldaker, G. B., and F. C. Conrad. "Estimation of Effect of Environmental Tobacco Smoke on Air Quality Within Passenger Cabins of Commercial Aircraft." *Environmental Science and Technology* 21 (October 1987): 994–999. Discussion, 22 (October 1988): 1238–1240.

Popham, W. James, Lance Potter, Dileep Bal, Michael Johnson, Jacquolyn Duerr, and Valerie Quinn. "Do Anti-Smoking Media Campaigns Help Smokers Quit?" *Public Health Reports* 108 (July-August 1993): 510–513.

Ross, J. A., Elia Sterling, Chris Collett, and Norman E. Kjono. "Controlling Environmental Tobacco Smoke in Offices." *Heating/Piping/Air Conditioning* 68 (May 1996): 76–78+.

Ross, Susan. "Second-Hand Smoke: The Asbestos and Benzene of the Nineties." *Arizona State Law Journal* 25 (Fall 1993): 713–731.

Ryan, P. B., J. D. Spengler, and P. F. Halfpenny. "Sequential Box Models for Indoor Air Quality: Application to Airliner Cabin Air Quality." *Atmospheric Environment* 22, no. 6 (1988): 1031–1038.

Schein, David D. "Should Employers Restrict Smoking in the Workplace?" *Labor Law Journal* 38 (March 1987): 173–178.

Schwartz, Allison D. "Environmental Tobacco Smoke and Its Effect on Children: Controlling Smoking in the Home." *Boston College Environmental Affairs Law Review* 20, no. 1 (1993): 135–171.

Sculco, Thomas W. "Smokers' Rights Legislation: Should the State 'Butt Out' of the Workplace?" *Boston College Law Review* 37 (July 1992): 879–902.

"Second-Hand Tobacco Smoke: Pros and Cons." *Congressional Digest* 73 (May 1994): 129–160.

Short, Lara W. "The Legal Implications of Workplace Smoking Policies." *Employee Respnsibilities and Rights Journal* 5 (March 1992): 65–73.

"Should Cigarettes Be Outlawed?" *U.S. News and World Report* 116 (April 18, 1994): 32–36+.

Siegelman, Stanley. "The Right to Blow Smoke." *Business and Health* 9 (September 1991): 79–82+.

Stone, R. "Air Quality: Bad News on Second-Hand Smoke." *Science* 257 (July 31, 1992): 607.

"Symposium on Smokers' and Non-Smokers' Rights." *St. Louis University Public Law Review* 13, no. 2 (1994): 547–785.

Tessaro, Irene. "Readiness to Change Smoking Behavior in a Community Health Center Population." *Journal of Community Health* 22 (February 1997): 15–31.

U.S. Congress. House Committee on Agriculture. Subcommittee on Specialty Crops and Natural Resources. *Review of the Smoke Study: Hearing, July 21, 1993.* 103d Cong., 1st sess. Washington, DC: GPO, 1993, 207 pp.

U.S. Congress. House Committee on Energy and Commerce. Subcommittee on Health and the Environment. *Designation of Smoking Areas in Federal Buildings: Hearings, June 12 and 27, 1986, on H. R. 4488 and H. R. 4546, Bills to Restrict Smoking to Designated Areas in All Buildings or Building Sections Occupied by the U.S. Government.* 99th Cong. 2d. sess. Washington, DC: GPO, 1987, 874 pp.

———. *Environmental Tobacco Smoke: Hearing, July 21, 1993.* 103d. Cong., 1st sess. Washington, DC: GPO, 1993, 226 pp.

———. *Health Effects of Smokeless Tobacco: Hearing, November 29, 1994.* 103d. Cong., 2d sess. Washington, DC: GPO, 1995, 201 pp.

U.S. Congress. House Committee on Public Works and Transportation. Subcommittee on Aviation. *Airliner Cabin Air Quality: Hearing May 18, 1994.* 103d. Cong., 2d sess. Washington, DC: GPO, 1994, 516 pp.

———. *To Ban Smoking on Airline Aircraft: Hearing, June 22, 1989.* 101st Cong., 1st sess. Washington, DC: GPO, 1990, 467 pp.

———. *To Ban Smoking on Airline Aircraft: Hearing, October 7, 1987.* 10th Cong., 1st sess. Washington, DC: GPO, 1988, 379 pp.

U.S. Congress. House Committee on Public Works and Transportation. Subcommittee on Public Buildings and Grounds. *To Protect Smoking in Federal Buildings: Hearings, March 11 and April 22, 1993, on H. R. 881.* 103d Cong., 1st sess. Washington, DC: GPO, 1993, 567 pp.

U.S. Congress. Senate Committee on Environment and Public Works. Subcommittee on Clean Air and Nuclear Regulation. *Assessing the Effects of Environmental Tobacco Smoke: Hearing, May 11, 1994, on S. 262, a Bill to Require the Administrator of the Environmen-*

tal Protection Agency to Promulgate Guidelines for Instituting a Nonsmoking Policy in Buildings Owned or Leased by the Federal Government, and S. 1680, a Bill to Amend the Toxic Substances Control Act to Protect the Public from Health Hazards Caused by Exposure to Environmental Tobacco Smoke, and for Other Purposes. 103d Cong., 2d sess. Washington, DC: GPO, 1994, 599 pp.

U.S. Congress. Senate Committee on Governmental Affairs. Subcommittee on Civil Service, Post Office, and General Services. *Non-Smokers Rights Act of 1985: Hearings, September 30–October 2, 1985, on S. 1440, to Restrict Smoking to Designated Areas in All United States Government Buildings.* 99th Cong., 1st sess. Washington, DC: GPO, 1986, 494 pp.

U.S. Department of Health and Human Services. *The Health Consequences of Involuntary Smoking.* Rockville, MD: DHHS, Office on Smoking and Health, 1986.

U.S. Laws and Statutes. *Compilation of Selected Acts within the Jurisdiction of the Committee on Commerce: Drug, and Related Law, January 1995.* 104th Cong., 1st sess. Washington, DC: GPO, 1995, 477 pp.

U.S. National Institution of Health. National Cancer Institute. *Strategies to Control Tobacco Use in the United States: A Blueprint for Public Health Action in the 1990s.* NIH Publication No. 92-3316. Bethesda, MD: the Institute, 1991, 307 pp.

Valentich, Marg. "Social Work and the Development of a Smoke-Free Society." *Social Work* 39 (July 1994): 439–450.

Vaughn, Dennis H. "Smoking in the Workplace: A Management Perspective." *Employee Relations Law Journal* 18 (Summer 1992): 123–139.

———. "Smoking in the Workplace: A Management Perspective." *Employee Relations Law Journal* 14 (Winter 1988-1989): 359–386.

Vaughn, Michael S., and Rolando V. del Carmen. "Legal and Policy Issues from the Supreme Court's Decision on Smoking in Prisons." *Federal Probation* 57 (September 1993): 34–39.

———. "Smoke-Free Prisons: Policy Dilemmas and Constitutional Issues." *Journal of Criminal Justice* 21, no. 2 (1993): 151–171.

Wagenknecht, L. E., T. A. Manolio, Stephen Sidney, G. L. Burke, and N. J. Haley. "Environmental Tobacco Smoke Exposure as Determined by Cotinine in Black and White Young Adults: The CARDIA Study." *Environmental Research* 63 (October 1993): 39–46.

Warner, Daniel M. "'We Do Not Hire Smokers': May Employers Discriminate against Smokers?" *Employee Responsibilities and Rights Journal* 7 (June 1994): 129–140.

Wartenberg, Daniel, Rodney Ehrlich, and David Lilienfeld. "Environmental Tobacco Smoke and Childhood Asthma: Comparing Exposure Metrics Using Probability Plots." *Environmental Research* 64 (February 1994): 122–135.

Wells, B. J., and R. L. Roberts. "SLO Smoke: The Anatomy of a Powerful Local Anti-Tobacco Law." *Journal of Environmental Health* 55 (April 1993): 9–13.

Other Atmospheric Pollutants

Atkins, D. H. F., and D. S. Lee. "Indoor Concentrations of Ammonia and the Potential Contributions of Humans to Atmospheric Budgets." *Atmospheric Environment, Part A* 27A, no. 1 (January 1993): 1–7.

Borrazzo, J. E., C. I. Davidson, and M. J. Small. "Stochastic Simulation of Diurnal Variations of CO, NO, and NO_2 Concentrations in Occupied Residences." *Atmospheric Environment, Part B* 26B (September 1992): 369–377.

Cano-Ruiz, J. A., D. Kong, R. B. Balas, and W. W. Nazaroff. "Removal of Reactive Gases at Indoor Surfaces: Combining Mass Transport and Surface Kinetics." *Atmospheric Environment, Part A* 27A, no. 13 (September 1993): 2039–2050.

Chang, J. C. S., and K. A. Krebs. "Evaluation of Para-dichlorobenzene Emissions from Solid Moth Repellant as a Source of Indoor Air Pollution." *Journal of the Air and Waste Management Association* 42 (September 1992): 1214–1217.

Chuang, J. C., M. R. Kuhlman, and N. K. Wilson. "Evaluation of Methods for Simultaneous Collection and Determination of Nicotine and Polynuclear Aromatic Hydrocarbons in Indoor Air." *Environmental Science and Technology* 24 (May 1990): 661–665.

Chuang, J. C., G. A. Mack, M. R. Kuhlman, and N. K. Wilson. "Polycyclic Aromatic Hydrocarbons and Their Derivatives in Indoor and Outdoor Air in an Eight-Home Study." *Atmospheric Environment, Part B* 25B, no. 3 (1991): 369–380. Correction, 26B, no. 1 (March 1992): 141.

Conrad, F. "Surgical and Other Aerosols: Protection in the Operating Room." *Professional Safety* 39 (August 1994): 28–30.

Crawford, Robert J., Steven L. Cloutier, and David C. Rovell-Rixx. "Evaluation of OSHA Method 5 for Measuring Chloroform in Pulp and Paper Industry Workplace and Ambient Atmospheres." *American Industrial Hygiene Association Journal* 53 (March 1992): 210–215.

Edgerton, S. A., D. V. Kenny, and D. W. Joseph. "Determination of Amines in Indoor Air from Steam Humidification." *Environmental Science and Technology* 23 (April 1989): 484–488.

Estill, C. F., and A. B. Spencer. "Case Study: Control of Methylene Chloride Exposure During Furniture Stripping." *American Industrial Hygiene Association Journal* 57 (January 1996): 43–49. Discussion, 57 (June 1996): 575–576.

Fernandez-Bremauntz, A. A., and M. R. Ashmore. "Exposure of Commuters to Carbon Monoxide in Mexico City—I. Measurement of In-Vehicle Concentrations." *Atmospheric Environment* (England) 29, no. 4 (March 1995): 525–532.

Graff, Gordon. "The Chlorine Controversy." *Technology Review* 98 (January 1995): 54–60.

Gulyas, H., and L. Hemmerling. "Tetrachloroethene Air Pollution from Coin-Operated Dry Cleaning Establishments." *Environmental Research* 53 (October 1990): 90–99.

Guo, Zhiski, Bruce A. Tichenor, Mark A. Mason, and C. Michelle Plunket. "The Temperature Dependence of the Emissions of Perchloroethylene from Dry Cleaned Fabrics." *Environmental Research* 52 (June 1990): 107–115.

Hamilton, K., M. F. Laliberte, R. Heslegrave, and S. Khan. "Visual/Vestibular Effects of Inert Gas Narcosis." *Ergonomics* 36 (August 1993): 891–898.

Harper, M. "A Novel Sampling Method for Methyl Methacrylate in Work-Place Air." *American Industrial Hygiene Association Journal* 53 (December 1992): 773–775.

Highsmith, V. R., D. L. Costa, M. S. Germani, and R. J. Hardy. "Physical and Chemical Characterization of Indoor Aerosols Resulting from Use of Tap Water in Portable Home Humidifiers." *Environmental Science and Technology* 26 (April 1992): 673–680.

Hisham, M. W. M., and D. Grosjean. "Air Pollution in Southern California Museums: Indoor and Outdoor Levels of Nitrogen Dioxide, Peroxyacetyl Nitrate, Nitric Acid, and Chlorinated Hydrocarbons." *Environmental Science and Technology* 25 (May 1991): 857–862.

Hollander, Albert, Dick Heederik, Pieter Versloot, and Jeroen Douwes. "Inhibition and Enhancement in the Analysis of Airborne Endotoxin Levels in Various Occupational Environments." *American Industrial Hygiene Association Journal* 54 (November 1993): 647–653.

Kandpal, J. B., et al. "Comparison of CO, NO_2, and HCHO Emissions from Biomass Combustion in Traditional and Improved Cookstoves." *Energy* 19 (November 1994): 1151–1155.

Karlsson, E. "Indoor Deposition Reducing the Effect of Toxic Gas Clouds in Ordinary Buildings." *Journal of Hazardous Materials* 38 (August 1994): 313–327.

Karlsson, E., and U. Huber. "Influence of Desorption on the Indoor Concentration of Toxic Gases." *Journal of Hazardous Materials* 49 (July 1996): 15–27.

Keating, G. A., T. E. McKone, and J. W. Gillett. "Measured and Estimated Air Concentrations of Chloroform in Showers: Effects of Water Temperature and Aerosols." *Atmospheric Environment* (England) 31, no. 2 (January 1997): 123–130.

Koutrakis, Petros, Susan L. K. Briggs, and Brian P. Leaderer. "Source Apportionment of Indoor Aerosols in Suffolk and Onondaga Counties, New York." *Environmental Science and Technology* 26 (March 1992): 521–527.

Marbury, M. C., D. P. Harlos, J. M. Samet, and J. D. Spengler. "Indoor Residential NO_2 Concentrations in Albuquerque, New Mexico." *JAPCA* 38 (April 1988): 392–398.

Maseley, C. L., and M. R. Meyer. "Petroleum Contamination of an Elementary School: A Case History Involving Air, Soil-Gas, and Groundwater Monitoring." *Environmental Science and Technology* 26 (January 1992): 185–192.

McKone, T. E., and J. P. Knezovich. "The Transfer of Trichlorethylene (TCE) from a Shower to Indoor Air: Experimental Measurements and Their Implications." *Journal of the Air and Waste Management Association* 41 (March 1991): 282–286.

Myer, H. E., Sandra T. O'Block, and V. Dharmarajan. "A Survey of Airborne, HDI, HDI-Based Polyisocyanate, and Solvent Concentrations in the Manufacture and Application of Polyurethane Coatings." *American Industrial Hygiene Association Journal* 54 (November 1993): 663–670.

Nonnemacher, G. S., and S. I. Foster. "Hydrogen Fluoride Is Cut from Compounding Process." *Modern Plastics* 73 (January 1996): 87–88+.

Orlando, R. A., and Y. J. Lao. "An Assessment of Exposure to Cyclohexylamine Arising from Steam Humidification of Indoor Air." *Journal of Environmental Health* 56 (December 1993): 6–9.

Pendergrass, S. M. "An Approach for Estimating Workplace Exposure to Otoluidine, Aniline, and Nitrobenzene." *American Industrial Hygiene Association Journal* 55 (August 1994): 733–737.

Raynor, P. C., Steven Cooper, and David Leith. "Evaporation of Polydisperse Multicomponent Oil Droplets." *American Industrial Hygiene Association Journal* 57 (December 1996): 1128–1136.

Ryan, P. B., M. L. Soczek, J. D. Spengler, and I. H. Billick. "The Boston Residential NO_2 Characterization Study: Preliminary Evaluation of the Survey Methodology." *JAPCA* 38 (January 1988): 22–27.

Shepherd, J. L., R. L. Corsi, and Jeff Kemp. "Chloroform in Indoor Air and Wastewater: The Role of Residential Washing Machines." *Journal of the Air and Waste Management Association* 46 (July 1996): 631–642.

Spicer, C. W., D. W. Kenny, G. F. Ward, and I. H. Billick. "Transformations, Lifetimes, and Sources of NO_2, HONO, and HNO_3 in Indoor Environments." *Journal of the Air and Waste Management Association* 43 (November 1993): 1479–1485.

Tang, Y. Z. "Determination of C_1–C_4 Hydrocarbons in Air." *Analytical Chemistry* 65 (July 15, 1993): 1932–1935.

Tichenor, B. A., L. E. Sparks, M. D. Jackson, Zhishi Guo, M. A. Mason, C. M. Plunket, and S. A. Rasor. "Emissions of Pershlorethylene from Dry Cleaned Fabrics." *Atmospheric Environment, Part A* 24A, no. 5 (1990): 1219–1229.

Tidy, G., and J. N. Cape. "Ammonia Concentrations in Houses and Public Buildings." *Atmospheric Environment, Part A* 27A, no. 14 (October 1993): 2235–2237.

Wallace, L. A., et al. "The TEAM Study: Personal Exposures to Toxic Substances in Air, Drinking Water, and Breath of 400 Residents of New Jersey, North Carolina, and North Dakota." *Environmental Research* 43 (August 1987): 290–307.

Weschler, C. J., and H. C. Shields. "Production of the Hydroxyl Radical in Indoor Air." *Environmental Science and Technology* 30 (November 1996): 3250–3258.

Weschler, C. J., Helen C. Shields, and Datta V. Naik. "Indoor Chemistry Involving O_3, NO, and NO_2 as Evidenced by 14 Months of Measurements at a Site in Southern California." *Environmental Science and Technology* 28 (November 1994): 2120–2132.

Fuel Pollution

Butterfield, Patricia, Eldon Edmundson, Gerald LaCava, and June Penner. "Woodstoves and Indoor Air: The Effects on Preschoolers' Upper Respiratory Systems." *Journal of Environmental Health* 52 (November-December 1989): 172–173.

Crittenden, B. D., and E. M. H. Khater. "Fouling from Vaporizing Kerosene." *Journal of Heat/Transfer* 109 (August 1987): 583–589.

Kandpal, J. B., et al. "Indoor Air Pollution from Combustion of Wood and Dung Cake and Their Processed Fuels in Domestic Cookstoves." *Energy Conversion and Management* 36 (November 1995): 1073–1079.

———. "Indoor Air Pollution from Domestic Cookstoves using Coal, Kerosene, and LPG." *Energy Conversion and Management* 36 (November 1995): 1067–1072.

Leaderer, B. P., Patricia M. Boone, and S. Katharine Hammond. "Total Particle, Sulfate, and Acidic Aerosol Emissions from Kerosene Space Heaters." *Environmental Science and Technology* 24 (June 1990): 908–912.

Meckler, M. "Indoor Air Quality vs. Energy Efficiency: Impact of New Ventilation Standards." *Consulting-Specifying Engineer* 4 (July 1988): 82–88.

Mumford, J. L., et al. "Indoor Air Pollutants from Unvented Kerosene Heater Emissions in Mobile Homes: Studies on Particles, Semivolatile Organics, Carbon Monoxide, and Mutagenicity." *Environmental Science and Technology* 25 (October 1991): 1732–1738.

O'Sullivan, P. "Energy and IAQ Can Be Complementary (Indoor Air Pollution)." *Heating/Piping/Air Conditioning* 61 (February 1989): 37–42.

Raiyani, C. V., et al. "Characterization and Problems of Indoor Pollution Due to Cooking Stove Smoke." *Atmospheric Environment, Part A* 27A, no. 11 (August 1993): 1643–1655.

Santamouris, M., et al. "Energy Characteristics and Savings Potential in Office Buildings." *Solar Energy* 52 (January 1994): 59–66.

Sauer, H. J. Jr., and R. H. Howell. "Examining the Indoor Air Quality and Energy Performance of VAV Systems." *ASHRAE Journal* 34 (July 1992): 43–50.

Scofield, C. M. "California Classroom VAV with IAQ and Energy Savings, Too." *Heating/Piping/Air Conditioning* 66 (January 1994): 89–93.

Stubington, J. F., et al. "Emissions and Efficiency from Production Cooktop Burners Firing Natural Gas." *Journal of the Institute of Energy* 67 (December 1994): 143–155.

Traynor, G. W., M. G. Apte, A. R. Carruthers, J. F. Dillworth, R. J. Prill, D. T. Grimsrud, and B. H. Turk. "The Effects of Infiltration and Insulation on the Source Strengths and Indoor Air Pollution from Combustion Space Heating Appliances." *JAPCA* 38 (August 1988): 1011–1015.

Traynor, G. W., M. G. Apte, A. R. Carruthers, J. F. Dillworth, D. T. Grimsrud, and L.A. Gundel. "Indoor Air Pollution Due to Emissions from Wood-Burning Stoves." *Environmental Science and Technology* 21 (July 1987): 691–697.

Traynor, G. W., M. G. Apte, H. A. Sokol, J. C. Chuang, W. G. Tucker, and J. L. Mumford. "Selected Organic Pollutant Emissions from Unvented Kerosene Space Heaters." *Environmental Science and Technology* 24 (August 1990): 1265–1270.

U.S. Congress. House Committee on Education and Labor. Subcommittee on Health and Safety. *Oversight Hearings to Review Extent to Which Exposure to Fumes from Diesel Engines in Closed Work Places Are Affecting the Health and Safety of Workers: Hearings July 12–August 2, 1989.* 101st Cong., 1st sess. Washington, DC: GPO, 1989, 353 pp.

Physical Pollutants

Radon

Bierma, T. J., K. G. Croke, and Daniel Swartzman. "Accuracy and Precision of Home Radon Monitoring and the Effectiveness of EPA Monitoring Guidelines." *JAPCA* 39 (July 1989): 953–959.

Billings, C. H. "Residential Ventilation for Removal of Radon." *Public Works* 127 (December 1996): 72–75.

Hopke, P. K., et al. "Assessment of the Exposure to and Dose from Radon Decay Products in Normally Occupied Homes." *Environmental Science and Technology* 29 (May 1995): 1359–1364.

Page, S. "EPA's Strategy to Reduce Risk of Radon (Indoor Radon Abatement Act of 1988 [IRAA])." *Journal of Environmental Health* 56 (December 1993): 27–36.

Pushen, J. S., and C. B. Nelson. "EPA's Perspective on Risks from Residential Radon Exposure." *Journal of the Air Pollution Control Association* 39, no. 7 (1989): 915–920.

Scheberle, Denise. "Radon and Asbestos: A Study of Agenda Setting and Casual Stories." *Policy Studies Journal* 22 (Spring 1994): 74–86.

Radon Pollution

Alter, H. W., and R. A. Oswald. "Nationwide Distribution of Indoor Radon Measurements: A Preliminary Data Base." *JAPCA* 37 (March 1987): 227–231.

Bolch, Ben, and Harold Lyons. "A Multibillion-Dollar Radon Scare." *Public Interest* (Washington) (Spring 1990): 61–67.

"EPA Urges Schools to Check for Radon." *JAPCA* 39 (June 1989): 878–882.

Fisher, Ann, G. H. McClelland, W. D. Schulze, and J. K. Doyle. "Communicating the Risk from Radon." *Journal of the Air and Waste Management Association* 41 (November 1991): 1440–1445.

Gammage, R. B., C. D. Dudney, D. L. Wilson, R. J. Saultz, and B. C. Bauer. "Subterranean Transport of Radon and Elevated Indoor Radon in Hilly Karst Terrains." *Atmospheric Environment, Part A* 26A, no. 12 (August 1992): 2237–2246.

Harley, N. H., and J. H. Harley. "Indoor Radon: A Natural Risk." *Nuclear Safety* 32 (October-December 1991): 537–543.

"Indoor Air: No Longer Seen as a Safe Haven, Air Indoors Presents Special Pollution Problems." *EPA Journal* 19 (October-December 1993): 6–39.

Kutzmark, Tammy, and Donald Geis. "The Radon Problem: A Solution Is Easier Than You Think. *Public Management* 76 (May 1994): 6–15.

Mose, D. G., G. W. Mushrush, and C. E. Chrosniak. "Soil Radon, Permeability, and Indoor Prediction." *Environmental Geology* 19 (March-April 1992): 91–96.

Nazaroff, W. W., S. R. Lewis, S. M. Doyle, B. A. Moed, and A. V. Nero. "Experiments on Pollutant Transport from Soil into Residential Basements by Pressure-Driven Airflow." *Environmental Science and Technology* 21 (May 1987): 459–466.

Nero, Anthony V. Jr. "A National Strategy for Indoor Radon." *Issues in Science and Technology* 9 (Fall 1992): 33–40.

New Jersey Department of Environmental Protection. Bureau of Environmental Radiation. *Statewide Scientific Study of Radon: Summary Report: Task 7 Final Report*. 1989, various paging.

"Radon: Pinpointing a Mystery." *EPA Journal* 12 (August 1986): 2–15.

Schery, S. D., and S. Whittlestone. "Evidence of High Deposition of Ultrafine Particles at Mauna Loa Observatory." *Atmospheric Environment* (England) 29, no. 22 (November 1995): 3319–3324. Discussion, 30, no. 21 (November 1996): 3683–3685.

Sextro, R. G. "Understanding the Origin of Radon Indoors—Building a Predictive Capability." *Atmospheric Environment* 21, no. 2 (1987): 431–438.

Stone, R. "New Radon Study: No Smoking Gun." *Science* 263 (January 28, 1994): 465.

Treffer, Brough E. "Radon Gas May Seep into Your Liability." *Real Estate Appraiser and Analyst* 53 (Spring 1987): 21–24.

"The Unseen Threat." *Chemistry and Industry,* no. 19 (October 4, 1993): 736.

Whittlestone, S., E. Robinson, and S. Ryan. "Radon at the Mauna Loa Observatory: Transport from Distant Continents." *Atmospheric Environment* 26A, no. 2 (1992): 251–260.

Radon Control

Bocanegra, R., and P. K. Hopke. "Theoretical Evaluation of Indoor Radon Control Using a Carbon Adsorption System." *JAPCA* 39 (March 1989): 305–309.

Bonnefous, Y. C., A. J. Gadgll, W. J. Fisk, R. J. Prill, and A. R. Nematollahl. "Field Study and Numerical Simulation of Substab Ventilation Systems." *Environmental Science and Technology* 26 (September 1992): 1752–1758.

Cavallo, A., K. Gadsby, and T. A. Reddy. "Use of Natural Basement Ventilation to Control Radon in Single Family Dwellings." *Atmospheric Environment, Part A* 26A, no. 12 (August 1992): 2251–2256.

Christian, J. E. "Energy Efficient Residential Building Foundations." *ASHRAE Journal* 33 (November 1991): 36+.

Clarkin, M., and T. Brennan. *Radon-Resistant Construction Techniques for New Residential Construction.* EPA/625/2-91/032. Washington, DC: EPA, Office of Research and Development, 1991.

Evans, J. S., N. C. Hawkins, and J. D. Graham. "The Value of Monitoring for Radon in the Home: A Decision Analysis." *JAPCA* 38 (November 1988): 1380–1385.

Fisher, Ann, et al. "Schools Respond to Risk Management Programs for Asbestos, Lead in Drinking Water and Radon." *Risk: Issues in Health and Safety* 4 (Fall 1993): 309–328.

Garbesi, Karina., Richard G. Sextro, W. J. Fisk, M. P. Modera, and K. L. Revzan. "Soil-Gas Entry into an Experimental Basement: Model Measurement Comparisons and Seasonal Effects." *Environmental Science and Technology* 27 (March 1993): 466–473.

Hamilton, M. A. "Radon Reduction: A Three-Year Folow-up." *Journal of Environmental Health* 56 (December 1993): 19–21.

Holford, D. J., and H. D. Freeman. "Effectiveness of a Passive Subslab Ventilation System in Reducing Radon Concentration in a Home." *Environmental Science and Technology* 30 (October 1996): 2914–2920.

Leonard, B. E. "Ventilation Rates by Measurement of Induced Radon Time-Dependent Behavior—Theory, Application, and Evaluation." *Nuclear Technology* 104 (October 1993): 89–105.

Marcinowski, F., and S. Napolitano. "Reducing the Risks from Radon." *Journal of the Air and Waste Management Association* 43 (July 1993): 955–962.

Neher, P. "Radon Monitor." *Electronics Now* 65 (February 1994): 66–70.

———. "Radon Monitor." *Electronics Now* 65 (January 1994): 56–62. Discussion, 65 (July 1994): 16.

Page, S. "EPA's Strategy to Reduce Risk of Radon (Indoor Radon Abatement Act of 1988 [IRAA])." *Journal of Environmental Health* 56 (December 1993): 27–36.

Pirrone, N., and S. A. Batterman. "Cost-Effective Strategies to Control Radon in Residences." *Journal of Environmental Engineering* 121 (February 1995): 120–131.

Renken, K. J., and S. J. Konopacki. "An Innovative Radon Mitigation-Energy Conservation Retrofit System for Residential Buildings." *Journal of the Air and Waste Management Association* 43 (March 1993): 310–315.

Sheets, R. W. "Structural Dependence of Diurnal Fluctuations of Radon Progeny in Residential Buildings." *Journal of the Air and Waste Management Association* 42 (April 1992): 457–459.

"Subcommittee of E-6 on Performance of Buildings Developing Standards for Radon Mitigation." *ASTM Standardization News* 20 (July 1992): 20.

Van Netten, C., R. B. Brands, D. R. Morley, and B. E. Sabels. "Development of a Long-Term Personal Radon Monitor." *American Industrial Hygiene Association Journal* 56 (November 1995): 1107–1110.

"What Price Safety?" *Chemistry and Industry,* no. 15 (August 1, 1994): 586.

Radon Concentration

Brager, G. S., A. V. Nero, and C. L. Tien. "Transport and Deposition of Indoor Radon Decay Products." *Atmospheric Environment, Part B* 25B, no. 3 (1991): 343–368.

Brown, C. E., et al. "Statistical Analyses of the Radon-222 Potential of Rocks in Virginia, U.S.A." *Environmental Geology and Water Sciences* 19 (May/June 1992): 193–202.

Cohen, B. L. "Compilation and Integration of Studies of Radon Levels in U.S. Homes by States and Counties." *Critical Reviews in Environmental Control* 22, nos. 3–4 (1992): 243–364.

Ennemoser, O., W. Ambach, P. Brunner, P. Schneider, W. Oberaigner, F. Purtscheller, and V. Stingl. "Unusually High Indoor Radon Concentrations." *Atmospheric Environment, Part A* 27A, no. 14 (1993): 2169–2172.

"EPA, PHS Call for Radon Testing in Most U.S. Homes." *JAPCA* 38 (November 1938): 1450–1453.

Keller, G., H. Schneiders, M. Schütz, A. Siehl, and R. Stamm. "Indoor Radon Correlated with Soil and Subsoil Radon Potential—A Case Study." *Environmental Geology and Water Sciences* 19 (March/April 1992): 113–119.

Kodosky, L. G. "An Evaluation of Residential Air Radon Concentrations and Related Variables in Southeast Michigan." *Environmental Geology* 23 (February 1994): 65–72.

Lowry, J. D., and S. B. Lowry. "Radionuclides in Drinking Water." *American Water Works Association Journal* 80 (July 1988): 50–64.

Nero, A. V., M. B. Schwehr, W. W. Nazaroff, and K. L. Revzan. "Distribution of Airborne Radon-222 Concentrations in U.S. Homes." *Science* 234 (November 21, 1986): 992–997.

Riley, J. A. "Student Measurement of Radon Gas Concentrations." *American Journal of Physics* 64 (January 1996): 72–77.

Riley, W. J., A. J. Gadgil, Y. C. Bonnefous, and W. W. Nazaroff. "The Effect of Steady Winds on Radon-222 Entry from Soil into Houses." *Atmospheric Environment* (England) 30, no. 7 (April 1996): 1167–1176.

Schery, S. D., and S. Whittlestone. "Evidence of High Deposition of Ultrafine Particles at Mauna Loa Observatory." *Atmospheric Environment* (England) 29, no. 22 (November 1995): 3319–3324.

U.S. Environmental Protection Agency. Office of Radiation Programs. *Radon Measurement in Schools: An Interim Report.* EPA-520/1-89-010. Washington, DC: EPA, 1989, 33 pp.

Health Risks

Cohen, B. L., and G. A. Colditz. "Tests of the Linear No Threshold Theory for Lung Cancer Induced by Exposure to Radon." *Environmental Research* 64 (January 1994): 65–89.

Conrath, S. M., and L. Kolb. "The Health Risk of Radon." *Journal of Environmental Health* 58 (October 1995): 24–26.

Crawford, W. A. "On Air Pollution, Environmental Tobacco Smoke, Radon, and Lung Cancer." *JAPCA* 38 (November 1988): 386–391.

Ferng, S.-F., and J. K. Lawson. "Residents in a High Radon Potential Geographic Area: Their Risk Perception and Attitude toward Testing and Mitigation." *Journal of Environmental Health* 58 (January-February 1996): 13–17.

Halpern, M., and K. E. Warner. "Radon Risk Perception and Testing: Sociodemographic Correlates." *Journal of Environmental Health* 56 (March 1994): 31–35.

Hopke, P. K., Bent Jensen, Chic-Shan Li, Nathan Montassier, Pietr Wasiolek, A. J. Cavallo, Kenneth Gatsby, Robert H. Socolow, and A. C. James. "Assessment of the Exposure to and Dose from Radon Decay Products in Normally Occupied Homes." *Environmental Science and Technology* 29 (May 1995): 1359–1364.

Horgan, J. "Radon's Risks." *Scientific American* 271 (August 1994): 14+.

Nero, A. V. "Estimated Risk of Lung Cancer from Exposure to Radon Decay Products in U.S. Homes: A Brief Review." *Atmospheric Environment* 22, no. 10 (1985): 2205–2211.

Nichols, Rita M. "Construction Contractors Confront the Indoor Radon Hazard: Homeowners' Private Causes of Action and a Federal Response with the Indoor Radon Abatement Bill." *Washington University Journal of Urban and Contemporary Law* 37 (Spring 1990): 135–167.

Puskin, J. S., and C. B. Nelson. "EPA's Perspective on Risks from Residential Radon Exposure." *JAPCA* 39 (July 1989): 915–920.

Stiefer, P. S., and B. R. Weir. "Health Risk Attributable to Environmental Exposure: Radon." *Journal of Hazardous Materials* 39 (November 1994): 211–223.

U.S. General Accounting Office. *Air Pollution: Hazards of Radon Could Pose a National Health Problem: Report to the Pennsylvania Congressional Delegation, House of Representatives.* GAO/RCEP-86-170. Washington, DC: GAO, 1986, 55 pp.

Vonstille, W. T., and H. L. A. Sacarello. "Radon and Cancer: Florida Study Finds No Evidence of Increased Risks." *Journal of Environmental Health* 55 (November-December 1990): 25–28.

Weinstein, Neil D., et al. "Promoting Remedial Response to the Risk of Radon: Are Information Campaigns Enough?" *Science, Technology, and Human Values* 14 (Autumn 1989): 360–379.

Governmental Policies

Abelson, Philip H. "Radon Today: The Role of Flimflam in Public Policy." *Regulation* (Cato Institute) 14 (Fall 1991): 95–100.

Cross, Frank B., and Paula C. Murray. "Liability for Toxic Radon Gas in Residential Home Sales." *North Carolina Law Review* 66 (April 1988): 687–739.

"Environmental Issues/Radon." *Forum for Applied Research and Public Policy* 4 (Spring 1989): 4–25 (four articles).

Jackowitz, Anne Rickard. "Radon's Radioactive Ramifications: How Federal and State Governments Should Address the Problem." *Boston College Environmental Affairs Law Review* 16 (Winter 1988): 329–382.

Nazaroff, W. W., and K. Teichman. "Indoor Radon: Exploring U.S. Federal Policy for Controlling Human Exposures." *Environmental Science and Technology* 24 (June 1990): 774–782.

New York (State). Joint Legislative Commission. *Radon: Risk, Reality, and Reason.* Albany, NY: the Commission, 1990, 56 pp.

Nichols, Rita M. "Construction Contractors Confront the Indoor Radon Hazard: Homeowners' Private Causes of Action and a Federal Response with the Indoor Radon Abatement Bill." *Washington University Journal of Urban and Contemporary Law* 37 (Spring 1990): 135–167.

Scheberle, Denise. "Radon and Asbestos: A Study of Agenda Setting and Casual Stories." *Policy Study Journal* 22 (Spring 1994): 74–86.

U.S. Congress. House Committee on Energy and Commerce. Subcommittee on Health and the Environment. *Indoor Air Pollution in Schools: Hearing, March 18, 1993.* 103d Cong., 1st sess. Washington, DC: GPO, 1993, 73 pp.

U.S. Congress. House Committee on Energy and Commerce. Subcommittee on Transportation and Hazardous Materials. *Radon Awareness and Disclosure: Hearing, June 3, 1992, on H. R. 3258, a Bill to Improve the Accuracy of Radon Testing Products and Services, to Increase Testing for Radon in Schools, to Create a Commission to Provide*

Increased Public Awareness of Radon, and for Other Purposes. 102d Cong., 2d sess. Washington, DC: GPO, 1992, 158 pp.

U.S. Congress. House Committee on Energy and Commerce. Subcommittee on Transportation, Tourism, and Hazardous Materials. *Radon Pollution Control Act of 1987: Hearing, April 23, 1987.* 100th Cong., 1st sess. Washington, DC: GPO, 1987, 119 pp.

U.S. Congress. House Committee on Science, Space, and Technology. Subcommittee on Natural Resources, Agriculture Research, and Environment. *Federal Efforts to Promote Radon Testing: Hearing, May 16, 1990.* 101st Cong., 2d sess. Washington, DC: GPO, 1990, 230 pp.

———. *Indoor Air Quality Act of 1988: Hearing, September 28, 1988.* 100th Cong., 2d sess. Washington, DC: GPO, 1989, 266 pp.

U.S. Congress. House Committee on Science and Technology. Subcommittee on Natural Resources, Agriculture Research, and Environment. *Radon and Indoor Air Pollution: Hearing, October 10, 1985.* 99th Cong., 1st sess. Washington, DC: GPO, 1986, 291 pp.

———. *Residential Radon Contamination and Indoor Quality Research Needs: Hearing, September 17, 1986.* 99th Cong., 2d sess. Washington, DC: GPO, 1987, 333 pp.

U.S. Congress. Senate Committee on Environment and Public Works. *Radon Contamination Problems in North Dakota: Hearing, August 20, 1987.* 100th Cong., 1st sess. Washington, DC: GPO, 1987, 64 pp.

———. *Radon Gas Issues: Joint Hearings, April 2, 1987, on S. 743 and S. 744, before the Subcommittees on Environmental Protection and Superfund and Environmental Oversight.* 100th Cong., 1st sess. Washington, DC: GPO, 1987, 124 pp.

U.S. Congress. Senate Committee on Environment and Public Works. Subcommittee on Environmental Protection. *Indoor Air Quality Act of 1987: Hearing, November 30, 1987, on S. 1629, a Bill to Authorize a National Program to Reduce the Threat to Human Health Posed by Exposure to Contaminants in the Air Indoors.* 100th Cong., 1st sess. Washington, DC: GPO, 1988, 146 pp.

U.S. Congress. Senate Committee on Environment and Public Works. Subcommittee on Superfund and Environmental Oversight. *Radon Contamination: How Federal Agencies Deal with It: Hearing, May 18, 1988, to Review a Report from the General Accounting Office.* 100th Cong., 2d sess. Washington, DC: GPO, 1988, 157 pp.

U.S. Congress. Senate Committee on Environment and Public Works. Subcommittee on Superfund, Ocean, and Water Protection. *Radon Testing for Safe Schools Act: Hearing, May 23, 1990, on S. 1697, a Bill to Require Local Educational Agencies to Conduct Testing for Radon Contamination in Schools, and for Other Purposes.* 101st Cong., 2d sess. Washington, DC: GPO, 1990, 163 pp.

U.S. General Accounting Office. *Air Pollution: Actions to Promote Radon Testing Report to the Committee on Science, Space, and Technology, House of Representatives.* Gaithersburg, MD: GAO 1992, 35 pp.

Asbestos

Health Risks

Alleman, James E., and Brooke T. Mossman. "Asbestos Revisited." *Scientific American* 277 (July 1997): 70–75.

Brookins, Douglas G., and Judith Binder. "The Great Asbestos Scam: A Fibrous Family of Minerals Is Under Atack, Even Though Its Most Important Member Proves to Be Low Risk." *World and I* 4 (December 1989): 336–343.

Cary, Peter. "The Asbestos Panic Attack: How the Feds Got Schools to Spend Billions on a Program That Really Didn't Amount to Much." *U.S. News and World Report* 118 (February 10, 1995): 61–63.

Corn, Jacqueline Karnell, and Morton Corn. "Changing Approaches to Assessment of Environmental Inhalation Risk: A Case Study." *Milbank Quarterly* 73, no. 1 (1995): 97–119.

Curran, P. T., and P. D. Salvador. "Taking the Emergency out of Asbestos." *Heating/Piping/Air Conditioning* 64 (March 1992): 59–63.

Dodson, Ronald F., et al. "Asbestos: Major Health Threat or Exaggerated Issue?" *Forum for Applied Research and Public Policy* 5 (Winter 1990): 67–75.

Fisher, Ann, et al. "Schools Respond to Risk Management Programs for Asbestos, Lead in Drinking Water, and Radon." *Risk: Issues in Health and Safety* 4 (Fall 1993): 309–328.

Grandjean, P., and E. Bach. "Indirect Exposures: The Significance of Bystanders at Work and at Home." *American Industrial Hygiene Association Journal* 47 (December 1986): 819–824.

Hedgecock, G. A. "The Identification and Control of Health Hazards." *Glass Technology* 28 (April 1987): 69–73.

Lave, L. B. "Health and Safety Risk Analysis: Information for Better Decisions." *Science* 236 (April 17, 1987): 291–295.

Marshall, Patrick G. "Asbestos: Are the Risks Acceptable? The Environmental Protection Agency has Labeled Asbestos a Public-Health Hazard, and Congress has Mandated That the Nation's Schools Clean Up Hazardous Asbestos: Some Members Now Want to Extend the Asbestos Regulations to Other Commercial and Public Buildings." *Editorial Research Reports* (May 9, 1990): 126–139.

Rendahl, Frederick. "Asbestos Risk." *Urban Land* 46 (July 1987): 18–21.

Webb, J. "Tragic Asbestos Error Will Kill Thousands." *New Scientist* 145, no. 4 (March 11, 1995): 4.

Webber, James S., Samuel Syrotynski, and Murray V. King. "Asbestos-Contaminated Drinking Water: Its Impact on Household Air." *Environmental Research* 46 (August 1988): 153–167.

Liability

Calhoun, Craig, and Henry K. Hiller. "Coping with Insidious Injuries: The Case of Johns-Manville Corporation and Asbestos Exposure." *Social Problems* (Society for the Study of Social Problems) 35 (April 1988): 162–181.

Cary, Peter. "The Asbestos Panic Attack: How the Feds Got Schools to Spend Billions on a Program That Really Didn't Amount to Much." *U.S. News and World Report* 118 (February 10, 1995): 61–63.

Edley, Christopher F. Jr., and Paul C. Weiler. "Asbestos: A Multi-Million-Dollar Crisis." *Harvard Journal on Legislation* 30 (Summer 1993): 383–408.

Olson, Kristin. *Legal Aspects of Asbestos Abatement: Responses to the Threat of Asbestos Containing Materials in School Buildings.* Topeka, KS: National Organization on Legal Problems of Education, 1986, 27 pp.

U.S. Congress. House Committee on the Judiciary. Subcommittee on Intellectual Property and Judicial Administration. *Asbestos Litigation Crisis in Federal and State Courts: Hearings, October 24, 1991–February 27, 1992.* 102d Cong., 1st and 2d sess. Washington, DC: GPO, 1993, 538 pp.

Removal

"Asbestos in Schools—A Special Report." *School Business Affairs* (December 1988): 26–42.

"Asbestos in the Workplace." *Public Works* 126 (April 15, 1995): A42-A48.

Bateson, C. "Results of a Survey Designed to Determine the Scientific and Economic Benefits of Asbestos Abatement." *American Industrial Hygiene Association Journal* 53 (June 1992): 381–386.

Bookspan, Shelley. "Asbestos—a Major Real Estate Concern." *Real Estate Law Journal* 19 (Fall 1990): 158–164.

Brown, S. K. "Asbestos Exposure During Renovation and Demolition of Asbestos—Cement Clad Buildings." *American Industrial Hygiene Association Journal* 48 (May 1987): 478–486.

Fischben, Alf, et al. "Respiratory Findings Among Millwright and Machinery Erectors: Identification of Health Hazards from Asbestos in Place at Work." *Environmental Research* 61 (April 1993): 25–35.

Ganick, N. "In-Place Management of Asbestos." *Heating/Piping/Air Conditioning* 64 (November 1992): 63–67.

Hardy, J. A., and A. E. Aust. "Iron in Asbestos Chemistry and Carcinogenicity." *Chemical Reviews* 95 (January-February 1995): 97–118.

"In-Place Management of Asbestos Increasing: Experts Say Awareness Still Is a Problem." *National Real Estate Investor* 32 (November 1990): 64+.

Keyes, D. L., et al. "Exposure to Airborne Asbestos Associated with Simulated Cable Installation Above a Suspended Ceiling." *American Industrial Hygiene Association Journal* 52 (November 1991): 479–484.

Kiefer, M. J., R. M. Buchan, T. J. Keefe, and K. D. Blehm. "A Predictive Model for Determining Asbestos Concentrations for Fibers Less Than Five Micrometers in Length." *Environmental Research* 43 (June 1987): 31–38.

Lee, R. J., D. R. Van Orden, and G. R. Dunmyre. "Interlaboratory Evaluation of the Breakage of Asbestos-Containing Dust Particles by Ultrasonic Agitation." *Environmental Science and Technology* 30 (October 1996): 3010–3015.

Lippy, J. D., and J. A. Boggs. "Measuring Airborne Asbestos." *Journal of Environmental Health* 52 (November-December 1989): 157–160.

Lundy, P., and M. Barer. "Asbestos-Containing Materials in New York City Buildings. *Environmental Research* 58 (June 1992): 15–24.

Phillips, C. C., and C. B. Hamilton. "A Preliminary Assessment of Asbestos Awareness and Control Measures in Brake and Clutch Repair Services in Knoxville and Knox County, Tennessee." *Journal of Environmental Health* 56 (April 1994): 7–12.

Reynolds, S. J., R. A. Kreiger, J. A. Bohn, Daniel Fish, Tim Marxhausen, and Charles McJilton. "Factors Affecting Airborne Concentrations of Asbestos in a Commercial Building." *American Industrial Hygiene Association Journal* 55 (September 1994): 823–828.

Roggli, V. L., M. H. George, and A. R. Brody. "Clearance and Dimensional Changes of Crocidolite Asbestos Fibers Isolated from Lungs of Rats Following Short-Term Exposure." *Environmental Research* 42 (February 1987): 94–105.

Ross, M. "The Schoolroom Asbestos Abatement Program: A Public Policy Debacle." *Environmental Geology* 26 (October 1995): 182–188.

Scheberle, Denise. "Radon and Asbestos: A Study of Agenda Setting and Casual Stories." *Policy Studies Journal* 22 (Spring 1994): 74–86.

Schneider, T., and J. Skotte. "Fiber Exposure Reassessed with the New Indices." *Environmental Research* 51 (February 1990): 108–116.

Skogstad, A., W. Eduard, and P. O. Huser. "A Laboratory Method for Generation of Replicate Filter Samples of Asbestos Fibers in Air." *American Industrial Hygiene Association Journal* 57 (August 1996): 741–745.

U.S. Environmental Protection Agency. *Guidance for Controlling Asbestos-Containing Materials in Buildings.* EPA 625/2-91/032. Washington, DC: EPA, Office of Pesticides and Toxic Substances, 1985.

U.S. Occupational Safety and Health Administration. *Asbestos Standards for the Construction Industry.* OSHA 3096. Washington, DC: OSHA, 1995, 39 pp.

Van Brunt, M. W. "Use Emergency Response Protocols to Limit Asbestos Dangers." *NFPA Journal* 85 (May/June 1992): 55–58+.

Welch, John F. "Asbestos in Buildings: An Overview for Community Associations." *Common Ground* (May-June 1987): 9–11.

Laws and Regulations

Baumann, David N. "More Asbestos Regulations: Are They Doing More Harm Than Good?" *Environmental Claims Journal* 2 (Spring 1990): 357–376.

Denson, F. A., and W. A. Onderick. "Managing Asbestos: Ten Costly Sins." *Power* 137 (January 1993): 46–48.

Gough, Michael. "Uncle Sam Flunks Asbestos Control in Schools: The EPA's Effort to Protect Students from Breathing Cancer-Causing Asbestos Could Actually Increase Their Risk." *Issues in Science and Technology* 4 (Spring 1988): 81–85.

Homburger, Thomas C. "Asbestos: Legal Issues and Answers." *Journal of Property Management* 53 (January-February 1988): 6–11.

Howitt, D. G., J. Hatfield, and G. Fishler. "The Difficulties with Low-Level Asbestos Exposure Assessments in Public, Commercial, and Industrial Buildings." *American Industrial Hygiene Association Journal* 54 (May 1993): 267–271.

Murray, Thomas H. "Regulating Asbestos Ethics, Politics, and Scientific Values." *Science, Technology, and Human Values* 11 (Summer 1986): 1–13.

Peck, Louis. "EPA Blues: If We Can't Regulate Asbestos, We Can't Regulate Anything." *Amicus Journal* 11 (Spring 1989): 18–21.

Rosenberg, Ernie, and John Wheeler. "Unreasonably at Risk: The EPA's Failure to Regulate Asbestos Under the Toxic Substances Control Act Raises Questions About the Law's Adequacy." *Environmental Forum* 10 (July-August 1993): 18–22.

U.S. Congress. House Committee on Energy and Commerce. Subcommittee on Commerce, Transportation, and Tourism. *Asbestos Exposure: Hearings, June 27, 1985, and March 4, 1986, on EPA Efforts to Control Asbestos Hazards and Asbestos Hazard Emergency Response Act of 1986.* 99th Cong., 1st and 2d sess. Washington, DC: GPO, 1986, 356 pp.

U.S. Congress. House Committee on Energy and Commerce. Subcommittee on Transportation, Tourism, and Hazardous Materials. *School Asbestos Abatement Program: Hearing, March 28, 1988, on H. R. 3893, a Bill to Amend Provisions of the Toxic Waste Substances Control Act Relating to Asbestos in the Nation's Schools by Providing*

Adequate Time for Local Educational Agencies to Submit Asbestos Management Plans to State Governors and to Begin Implementation of Those Plans. 100th Cong., 2d sess. Washington, DC: GPO, 1989, 265 pp.

U.S. Congress. House Committee on Government Operations. Subcommittee on Environment, Energy, and Natural Resources. *Asbestos Dangers: Presence in Schools and Incompetent Disposal: Hearing, August 3, 1987.* 100th Cong., 1st sess. Washington, DC: GPO, 1988, 378 pp.

———. *EPA's Administration of the Asbestos in Schools Program: Hearing, September 24, 1991.* 102d Cong., 1st sess. Washington, DC: GPO, 1992, 167 pp.

———. *EPA's Implementation of Laws Regulating Asbestos Hazards in Schools and in the Air: Hearing, June 1, 1988.* 100th Cong., 2d sess. Washington, DC: GPO, 1989, 230 pp.

U.S. Congress. Senate Committee on Environment and Public Works. Subcommittee on Hazardous Wastes and Toxic Substances. *Management of Asbestos in Public Buildings: Hearing, September 27, 1988, on S. 981, S. 1809, and S. 2687.* 100th Cong., 2d sess. Washington, DC: GPO, 1988, 226 pp.

U.S. Congress. Senate Committee on Environment and Public Works. Subcommittee on Toxic Substances and Environmental Oversight. *Hazardous Asbestos Abatement: Hearing, May 15, 1986, on S. 2083 and S. 2300, Bills Providing for the Abatement of Hazardous Asbestos.* 99th Cong., 2d sess. Washington, DC: GPO, 1986, 327 pp.

Lead

General

Bero, B. N., and M. C. von Braun. "The Effectiveness of the Freeze Fracture Carpet Grunding Technique in the Determination of Total Lead." *American Industrial Hygiene Association Journal* 57 (May 1996): 480–483.

Crouch, Keith, Taiyao Peng, and Donald J. Murdock. "Ventilation Control of Lead in Indoor Firing Ranges: Inlet Configuration and Booth and Fluctuating Flow Contributions." *American Industrial Hygiene Association Journal* 52 (February 1991): 81–91.

Eppler, R. A., and D. A. Eppler. "Formulating Lead-Free Glazes." *American Ceramic Society Bulletin* 75 (September 1996): 62–65.

Eskings, I., Z. Grabaric, and B. S. Grabaric. "Monitoring of Pyrocatechol Indoor Air Pollution." *Atmospheric Environment* (England) 29, no. 10 (May 1995): 1165–1170.

Ford, D. P., et al. "A Quantitative Approach to the Characterization of Cumulative and Average Solvent Exposure in Paint Manufacturing Plants." *American Industrial Hygiene Association Journal* 52 (June 1991): 226–234.

Jackson, P. R. "The Removal of Lead from Ceramic Tableware." *Glass Technology* 34 (June 1993): 98–101.

Lanphear, B. P., et al. "A Side-by-Side Comparison of Dust Collection Methods for Sampling Lead-Contaminated House Dust." *Environmental Research* 68 (February 1995): 114–123.

Moschandreas, D. J., and D. S. O'Dea. "Measurement of Perchloroethylene Indoor Air Levels Caused by Fugitive Emissions from Unvented Dry-to-Dry Dry Cleaning Units." *Journal of the Air and Waste Management Association* 45 (February 1995): 111–115.

Nair, Nathan, Amar Bhalla, and Rustum Roy. "Inorganic Lead Compounds in Electroceramics and Glasses." *American Ceramic Society Bulletin* 75 (January 1996): 77–82.

Pollack, S. "Solving the Lead Dilemma." *Technology Review* 92 (October 1989): 22–31.

Samuels, S. J., Carin Perkins, D. S. Sharp, Jon Rosenberg, and Linda Rudolph. "Design and Methods for a Survey of Lead Usage and Exposure Monitoring in California Industry." *American Industrial Hygiene Association Journal* 52 (October 1991): 403–408.

Schlecht, P. C., and J. H. Groff. "Environmental Lead Proficiency Analytical Testing (EL-PAT) Program." *American Industrial Hygiene Association Journal* 56 (October 1995): 1034–1040.

Tepper, J. S., et al. "Toxicological and Chemical Evaluation of Emissions from Carpet Samples." *American Industrial Hygiene Association Journal* 56 (February 1995): 158–170.

Vanter, M., et al. "Methods for Integrated Exposure Monitoring of Lead and Cadmium." *Environmental Research* 56 (October 1991): 78–89.

Wagner, B. S., and R. C. Diamond. "The Kansas City Warm Room Project: Economics, Energy Savings, and Health and Comfort Impacts." *Energy* 12 (June 1987): 447–457.

Zuckerman, J. "Decorating Tableware in the 90s." *American Ceramic Society Bulletin* 71 (September 1992): 1368–1370.

Health Risks

Brooks, W. "Lowering the Danger Standard: Will Lead Be Another Asbestos Racket?" *Recycling Today* 30 (May 1992): 28+.

Edison, M., and K. Tollestrup. "Blood Lead Levels and Remediation of an Abandoned Smelter Site." *Journal of Environmental Health* 57 (May 1995): 8–14.

Foulke, Judith E. "Lead Threat Lessens, But Mugs Pose Problems." *FDA Consumer* 27 (April 1993): 19–23.

Gerwel, Barbara, and Martha Stanbury, eds. *Adult Blood Lead Epidemiology and Surveillance in New Jersey, 1992–1993.* Trenton, NJ: New Jersey Department of Health, Division of Epidemiology, Environmental and Occupational Health Services, 1995, 26 pp.

Gurman, Lew. "Deadly Lead: Lead's Meance May Be Forgotten, But It Hasn't Disappeared." *Environmental Action* 17 (March-April 1986): 18–20.

Lauterback, S. "New Disposal Headache: PCBS in Lead-Covered Scrap Cable." *Electrical World* 210 (June 1996): 29–30.

Lipmann, M. "Lead and Human Health: Background and Recent Findings (1989 Alice Hamilton Lecture)." *Environmental Research* 51 (February 1990): 1–24.

Renner, R. "When Is Lead a Health Risk?" *Environmental Science and Technology* 29 (June 195): 256A–261A.

Schwartz, J., and R. Levin. "The Risk of Lead Toxicity in Homes with Lead Paint Hazard." *Environmental Research* (February 1991): 1–7.

U.S. Congress. House Committee on Ways and Means. Subcommittee on Select Revenue Measures. *Lead-Based Paint Hazard Abatement Act: Hearing, July 1, 1992, on H. R. 2922, to Amend the Public Health Service Act to Establish an Entitlement of States and Certain Political Subdivisions of States to Receive Grants for the Abatement of Health Hazards Associated with Lead-Based Paint, and to Amend the Internal Revenue Code of 1986 to Impose an Excise Tax and Establish a Trust Fund to Satisfy the Federal Obligations Arising from such Entitlement.* 102d Cong., 2d sess. Washington, DC: GPO, 1992, 213 pp.

Sources

Adgate, J. L., C. Weisel, Y. Wang, G. G. Rhoads, and P. J. Lioy. "Lead in House Dust: Relationships Between Exposure Metrics." *Environmental Research* 70 (August 1995): 134–147.

Farfel, M. R., and J. J. Chisolm Jr. "An Evaluation of Experimental Practices for Abatement of Residential Lead-Based Paint: Report on a Pilot Project." *Environmental Research* 55 (August 1991): 199–212.

Hunt, Andrew, D. L. Johnson, J. M. Watt, and Iain Thornton. "Characterizing the Sources of Particulate Lead in House Dust by Automated Scanning Electron Microscopy." *Environmental Science and Technology* 26 (August 1992): 1513–1523.

Children

Aschengrau, Ann, Alexa Beiser, David Bellinger, Donna Copenhafer, and Michael Weitzman. "The Impact of Soil Lead Abatement on Urban Children's Blood Lead Levels: Phase II Results from the Boston Lead-in-Soil Demonstration Project." *Environmental Research* 67 (November 1994): 125–148.

Bailey, Adrian J., James Sargent, David Goodman, Jean Freeman, and Mary Jean Brown. "Poisoned Landscapes: The Epidemiology of Environmental Lead Exposure in Massachusetts Children, 1990–1991." *Social Science and Medicine* 39, no. 6 (1994): 757–766.

Ballas, Christine, F. A. Rusczek, and Richard DiPentima. "Cost Benefit of Providing Comprehensive Case Management for Lead-Burdened Children." *Journal of Environmental Health* 57 (May 1995): 15–18.

Bernard, A. M., A. Vyskocil, H. Roels, J. Kriz, M. Kodl, and R. Lauwerys. "Renal Effects in Children Living in the Vicinity of a Lead Smelter." *Environmental Research* 68 (February 1995): 91–95.

Berney, Barbara. "Round and Round It Goes: The Epidemiology of Childhood Lead Poisoning, 1950–1990." *Milbank Quarterly* 71, no. 1 (1993): 3–39.

Center for Disease Control. Environmental Health Services Division. *Lead Poisoning in Children: A Problem in Your Community?* Rev. ed. Atlanta: Department of Health, Education, and Welfare. Public Health Services. Washington, DC: Center for Disease Control. Environmental Health Services Division, 1977, 8 pp.

Cooper, Mary H. "Lead Poisoning: Are Children Suffering Because of Weak Prevention Efforts?" *CQ Researcher* 2 (June 19, 1992): 527–547.

Cosgrove, E., P. McNulty, M. J. Brown, L. Okonski, P. Madigan, and J. Schmidt. "Childhood Lead Poisoning: Case Study Traces Source to Drinking Water." *Journal of Environmental Health* 52 (July-August 1989): 346–349.

Dugbatey, Kwesi, Gregory Evans, Marie T. Lienhop, and Margie Stelzer. "Community Partnerships in Preventing Childhood Lead Poisoning." *Journal of Environmental Health* 58 (November 1995): 6–10.

Florini, Karen L., and Ellen K. Silbergeld. "Getting the Lead Out: A Federal Effort to Address Lead-Based Paint Hazards in Low-Income Homes Can Protect Millions of Children from Brain Damage." *Science and Technology* 9 (Summer 1993): 33–39.

Fullmer, C. S. "Intestinal Calcium and Lead Absorption: Effects of Dietary Lead and Calcium." *Environmental Research* 54 (April 1991): 159–169.

Mushak, P. "Defining Lead as the Premiere Environmental Health Issue for Children in America: Criteria and Their Quantitative Application." *Environmental Research* 59 (December 1992): 281–309.

Mushak, P., and A. F. Crocetti. "Determination of Numbers of Lead-Exposed American Children as a Function of Lead Source: Integrated Summary of a Report to the U.S. Congress on Childhood Lead Poisoning." *Environmental Research* 50 (December 1989): 210–229.

Salkever, D. S. "Updated Estimates of Earnings Benefits from Reduced Exposure of Children to Environmental Lead." *Environmental Research* 70 (July 1995): 1–5.

Sutton, P. M., et al. "Lead Levels in the Household Enviroinment of Children in Three High-Risk Communities in California." *Environmental Research* 68 (January 1995): 45–57.

Liability

Abarbanel, Stephen J. "Insurance Coverage for Lead-Poisoning Claims: Who Pays?" *Environmental Claims Journal* 5 (Spring 1993): 431–441.

Pancak, Katherine A., et al. "Legal Duties of Property Owners Under Lead-Based Paint Laws." *Real Estate Law Journal* 24 (Summer 1995): 7–25.

Ponessa, Jeanne. "Government's Role in Cleaning Up Lead: The Biggest Problem Is Determining Who Will Pay the Bill: The Paint

Industry, Home Owners—or Government." *Governing* 5 (August 1992): 20–21.

Laws and Regulations

Brugge, Doug. "Market Share Legislation: Holding the Lead Pigment Companies Accountable for Their Role in Lead Poisoning." *New Solutions* 5 (Winter 1995): 74–80.

Labadorf, Allan, and R. Mark Keenan. "Lead Paint Removal: What Insurers Won't Tell You!" *Journal of Housing and Community Development* 52 (November-December 1995): 24–30.

Pancak, Katherine A., Thomas J. Miceli, and C. F. Sirmans. "Legal Duties of Property Owners Under Lead-Based Paint Laws." *Real Estate Law Journal* 24 (Summer 1995): 7–25.

Pollack, Stephanie. "Solving the Lead Dilemma: Stricter Regulations Are the Key to Developing Better Ways to Protect Children from Lead Paint in Homes." *Technology Review* 92 (October 1989): 22–31.

U.S. Congress. House Committee on Banking, Finance, and Urban Affairs. Subcommittee on Housing and Community Development. *Lead-Based Paint Poisoning Prevention Act: Hearing, September 22, 1988.* 100th Cong., 2d sess. Washington, DC: GPO, 1988, 223 pp.

U.S. Congress. House Committee on Energy and Commerce. Subcommittee on Health and the Environment. *Hazard of Lead in Schools and Day Care Facilities: Hearing, September 15, 1993.* 103d Cong., 1st sess. Washington, DC: GPO, 1993, 98 pp.

———. *Lead Contamination: Hearings, December 10, 1987, and July 13, 1988.* 100th Cong. Washington, DC: GPO, 1988, 677 pp.

———. *Lead Poisoning Hearings, April 25 and July 26, 1991, on H. R. 2840, a Bill to Amend the Public Health Service Act to Reduce Human Exposure to Lead in Residences, Schools for Young Children, and Day Care Centers, Including Exposure to Lead in Drinking Water.* 102d Cong., 1st sess. Washington, DC: GPO, 1991, 672 pp.

U.S. Congress. House Committee on Energy and Commerce. Subcommittee on Oversight and Investigations. *Lead in Housewares: Hearing, June 27, 1988.* 100th Cong., 2d sess. Washington, DC: GPO, 1988, 490 pp.

U.S. Congress. Senate Committee on Banking, Housing, and Urban Affairs. *Lead-Based Paint Hazard in American Housing: Hear-*

ing, October 17, 1991, to Expand Federal Support for Testing, Containment, and Abatement of Lead-Based Paint Hazards Both in Federally Assisted and Private Housing. 102d Cong., 1st sess. Washington, DC: GPO, 1992, 465 pp.

U.S. Congress. Senate Committee on Banking, Housing, and Urban Affairs. Subcommittee on Housing and Urban Affairs. *The Residential Lead-Based Paint Hazard Reduction Act of 1992: Hearing, March 19, 1992, on S. 2341, to Provide for the Assessment and Reduction of Lead-Based Paint Hazards in Housing*. 102d Cong., 2d sess. Washington, DC: GPO, 1992, 289 pp.

U.S. Congress. Senate Committee on Environment and Public Works. Subcommittee on Toxic Substances, Environmental Oversight, Research, and Development. *The Administration's Strategy to Reduce Lead Poisoning and Contamination Hearing, February 21, 1991*. 102d Cong., 1st sess. Washington, DC: GPO, 1991, 162 pp.

———. *Health Effects of Lead Exposure: Hearing, March 8, 1990*. 101st Cong., 2d sess. Washington, DC: GPO, 1990, 292 pp.

———. *The Lead Ban Act of 1990 and the Lead Exposure Reduction Act of 1990: Hearing, June 27, 1990, on S. 2593, a Bill to Reduce the Amount of Lead Contamination in the Environment, and S. 2637, a Bill to Amend the Toxic Substances Control Act to Reduce the Levels of Lead in the Environment and for Other Purposes*. 101st Cong., 2d sess. Washington, DC: GPO, 1990, 399 pp.

U.S. Congress. Senate Committee on Governmental Affairs. Ad Hoc Subcommittee on Consumer and Environmental Affairs. *Consumer Protection at the Food and Drug Administration: Hearing, September 27, 1991*. 102d Cong., 1st sess. Washington, DC: GPO, 1994, 145 pp.

———. *Lead in Ceramicware and Crystal: An Avoidable Risk: Hearing, March 27, 1992*. 102d Cong., 2d sess. Washington, DC: 1992, 234 pp.

U.S. Department of Housing and Urban Development. Office of Policy Development and Research. *The HUD Lead-Based Paint Abatement Demonstration*. HUD-1316-PDR. Washington, DC: HUD, 1991, 3 vols.

Heavy Metals

Altmayer, F. "Hexavalent Chromium PEL (Question and Answer)." *Plating and Surface Finishing* 83 (June 1996): 35–37.

Bonilla, J. V., and R. A. Milbrath. "Cadmium in Plastic Processing Fumes from Injection Molding." *American Industrial Hygiene Association Journal* 55 (November 1994): 1069–1071.

Linnainmaa, Markku, Juhani Kangas, and Pentti Kalliokoski. "Exposure to Airborne Metals in the Manufacture and Maintenance of Hard Metal and Stellite Blades." *American Industrial Hygiene Association Journal* 57 (February 1996): 196–201.

Naitove, M. H. "Cadmium Pigments No Hazard in Molding." *Plastics Technology* 39 (November 1993): 89.

Persson, D., and C. Leygraf. "Metal Carboxylate Formation During Indoor Atmospheric Corrosion of Cu, Zn, and Ni." *Journal of the Electrochemical Society* 142 (May 1995): 1468–1477.

Teschke, Kay, Stephen A. Marion, Marielle J. A. van Zuylen, and Susan M. Kennedy. "Maintenance of Stellite and Tungsten Carbide Saw Tips: Determinants of Exposure to Cobalt and Chromium." *American Industrial Hygiene Association Journal* 56 (July 1995): 661–669.

Particulates

Abdel-Kader, H. M., Roy J. Rando, and Yehia Y. Hammad. "Long-Term Cotton Dust Exposure in the Textile Industry." *American Industrial Hygiene Association Journal* 48 (June 1987): 545–550.

Ashley, Kevin, Thomas J. Fischbach, and Ruiguang Song. "Evaluation of a Chemical Spot-Test Kit for the Detection of Airborne Particulat Lead in the Workplace." *American Industrial Hygiene Association Journal* 57 (February 1996): 161–165.

Boatman, E. S., D. Covert, D. Kalman, D. Luchtel, and G. S. Omenn. "Physical, Morphological, and Chemical Studies of Dusts Derived from the Machining of Composite-Epoxy Materials." *Environmental Research* 45 (April 1988): 242–255.

Crook, B., J. F. Robertson, S. A. Travers Glass, E. M. Botheroyd, J. Lacey, and M. D. Topping. "Airborne Dust, Ammonia, Microorganisms, and Antigens in Pig Confinement Houses and the Respiratory Health of Exposed Farm Workers." *American Industrial Hygiene Association Journal* 52 (July 1991): 271–279.

Esmen, N. A. "Adhesion and Aerodynamic Resuspension of Fibrous Particles." *Journal of Environmental Engineering* 122 (May 1996): 379–383.

Jafrey, T. S. A. M. "Levels of Airborne Man-made Mineral Fibres in U.K. Dwellings." *Atmospheric Environment, Part A* 24A, no. 1 (1990): 133–146.

Ligocki, M. P., L. G. Salmon, Theresa Fall, M. C. Jones, W. W. Nazaroff, and G. R. Cass. "Characteristics of Airborne Particles Inside Southern California Museums." *Atmospheric Environment, Part A* 27A, no. 5 (April 1993): 697–711.

Lohr, V. I., and C. H. Pearson-Mims. "Particulate Matter Accumulation on Horizontal Surface in Interiors: Influence of Foliage Plants." *Atmospheric Environment* (England) 30, no. 14 (July 1996): 2565–2568.

Nazaroff, W. W., Lynn G. Salmon, and Glen R. Cass. "Concentration and Fate of Airborne Particles in Museums." *Environmental Science and Technology* 24 (January 1990): 66–77.

Nishioka, M. G., H. M. Burkholder, M. C. Brinkman, and S. M. Gordon. "Measuring Transport of Lawn-Applied Herbicide Acids from Turf to Home: Correlation of Dislodgeable 2, 4-D Turf Residues with Carpet Dust, and Carpet Surface Residues." *Environmental Science and Technology* 30 (November 1996): 3313–3320.

Ott, Wayne, Paul Switzer, and John Robinson. "Particle Concentrations inside a Tavern before and after Prohibition of Smoking: Evaluating the Performance of an Indoor Air Quality Model." *Journal of the Air and Waste Management Association* 46 (December 1996): 1120–1134.

Owen, M. K., D. S. Ensor, and L. E. Sparks. "Airborne Particle Sizes and Sources Found in Indoor Air." *Atmospheric Environment, Part A* 26A, no. 12 (August 1992): 2149–2162.

Reighard, T. S., and S. V. Olesik. "Comparison of Supercritical Fluids and Enhanced-Fluidity Liquids for the Extraction of Phenolic Pollutants from House Dust." *Analytical Chemistry* 68 (October 15, 1996): 3612–3621.

Thatcher, T. L., and D. W. Layton. "Deposition, Resuspension, and Penetration of Pesticides within a Residence." *Atmospheric Environment* (England) 29, no. 13 (July 1995): 1487–1497.

Van der Wal, J. F., R. Ebens, and J. Tempelman. "Man-made Mineral Fibres in Homes Caused by Therman Insulation." *Atmospheric Environment* 21, no. 1 (1987): 13–19. Correction, 21, no. 3 (1987): 731.

Van Houdt, J. J. "Mutagenic Activity of Airborne Particulate Matter in Indoor and Outdoor Environments." *Atmospheric Environment, Part B* 24B, no. 2 (1990): 207–220.

Wallace, L. "Indoor Particles: A Review." *Journal of the Air and Waste Management Association* 46 (February 1996): 98–126.

Bioaerosols

Bacteria

DeKoster, J. A., and P. S. Thorne. "Bioaerosol Concentrations in Noncompliant, Compliant, and Intervention Homes in the Midwest." *American Industrial Hygiene Association Journal* 56 (June 1995): 573–580.

Epstein, E. "Composting and Bioaerosols." *BioCycle* 35 (January 1994): 51–58.

Foo, S. C., J. Jeyaratnam, C. N. Ong, N.Y. Khoo, D. Koh, and S. E. Chia. "Biological Monitoring for Occupational Exposure to Toluene." *American Industrial Hygiene Association Journal* 52 (May 1991): 212–217.

Jensen, P. A., W. F. Todd, G. N. Davies, and P. V. Scarpino. "Evaluation of Eight Bioaerosol Samplers Challenged with Aerosols of Free Bacteria." *American Industrial Hygiene Association Journal* 53 (October 1992): 660–667.

Korsgaard, J. "Preventive Measures in House-Dust Allergy." *American Review of Respiratory Diseases* 125 (1982): 80–84.

McJilton, C. E., S. J. Reynolds, A. J. Streifel, and R. L. Pearson. "Bacteria and Indoor Odor Problems—Three Case Studies." *American Industrial Hygiene Association Journal* 51 (October 1990): 545–549.

Moschandreas, D. J., Daniel K. Cha, and Jon Qian. "Measurement of Indoor Bioaerosol Levels by a Direct Counting Method." *Journal of Environmental Engineering* 122 (May 1996): 374–378.

Rhodes, W. W., M. G. Rinaldi, and G. W. Gorman. "Reduction and Growth Inhibition of Microorganisms in Commercial and Institutional Environments." *Journal of Environmental Health* 58 (October 1995): 12–19.

Van der Werf, P. "Bioaerosols at a Canadian Composting Facility." *BioCycle* 37 (September 1996): 78–80+.

Legionnaires' Disease

Clark, M., and S. G. Mechaud. "Poison Theory: Nickel Carbonyl Poisoning Suspected as Cause of Legionnaires' Disease." *Newsweek* 88 (September 6, 1976): 34.

Cravens, G., and J. S. Marr. "Tracking Down the Epidemic: Legionnaires' Disease." *New York Times Magazine* (December 12, 1976): 32–33+.

Culliton, B. J. "Legion Fever: Postmortem on an Investigation That Failed." *Science* 194 (December 3, 1976): 1025–1027.

Skaliy, P., and H. V. McEachern. "Survival of Legionnaires' Disease Bacterium in Water." *Annals of Internal Medicine* 90 (1979): 662–663.

Fungi

Chang, J. C. S., K. K. Foarde, and D. W. Vanosdell. "Growth Evaluation of Fungi (Penicillium and Aspergillus spp.) on Ceiling Tiles." *Atmospheric Environment* (England) 29, no. 17 (September 1995): 2331–2337.

Hollomon, D. W., and K. J. Brent. "Fungicide Resistance—Can We Cope with It?" *Chemistry and Industry,* no. 6 (March 20, 1989): 177–182.

Miller, J. D., and J. C. Young. "The Use of Ergosterol to Measure Exposure to Fungal Propaqules in Indoor Air." *American Industrial Hygiene Association Journal* 58 (January 1997): 39–43.

Parat, S., H. Fricker-Hidalgo, A. Perdrix, D. Bemer, N. Pelissier, and R. Grillot. "Airborne Fungal Contamination in Air-Conditioning Systems: Effect of Filtering and Humidifying Devices." *American Industrial Hygiene Association Journal* 57 (November 1996): 996–1001.

Pasanen, A.-L. "Airborne Mesophilic Fungal Spores in Various Residential Environments." *Atmospheric Environment, Part A* 26A, no. 16 (November 1992): 2861–2868.

Pasanen, A.-L., H. Heinoneni Tanski, P. Kalliokoski, and M. J. Jantienen. "Fungal Microcolonies on Indoor Surfaces—an Explanation for the Base-Level Fungal Spore Counts in Indoor Air." *Atmospheric Environment, Part B* 26B (March 1992): 117–120.

Pasanen, A.-L., M. Niininen, P. Kalliokoski, A. Nevalainen, and M. J. Jantienen. "Airborne Cladosporium and Other Fungi in

Damp Versus Reference Residences." *Atmospheric Environment, Part B* 26B (March 1992): 121–124.

Rao, Carol Y., Harriet A. Burge, and John C. S. Chang. "Review of Quantitative Standards and Guidelines for Fungi in Indoor Air." *Journal of the Air and Waste Management Association* 46 (September 1996): 899–908.

Reynolds, S. J., A. J. Streifel, and C. E. McJilton. "Elevated Airborne Concentrations of Fungi in Residential and Office Environments." *American Industrial Hygiene Association Journal* 51 (November 1990): 601–604.

Rijckaert, G. "Exposure to Fungi in Modern Homes." *Allergy* 36 (1981): 277–279.

Viranyi, F. "Mildew Diseases: A Continued Challenge." *Chemistry and Industry,* no. 12 (June 18, 1990): 401–403.

Volatile Organic Compounds

Location of VOCs

Bukowski, J. A., M. G. Robson, B. T. Buckley, D. W. Russell, and L. W. Meyer. "Air Levels of Volatile Organic Compounds Following Indoor Application of an Emulsifiable Concentrate Insecticide." *Environmental Science and Technology* 30 (August 1996): 2543–2546.

Fellin, P., and R. Otson. "Assessment of the Influence of Climate Factors on Concentration Levels of Volatile Organic Compounds (VOCs) in Canadian Homes." *Atmospheric Environment* (England) 28, no. 22 (December 1994): 3581–3586.

Fischer, M. L., et al. "Factors Affecting Indoor Air Concentrations of Volatile Organic Compounds at a Site of Subsurface Gasoline Contamination." *Environmental Science and Technology* 30 (October 1996): 2948–2957.

Hawkins, N. C., A. E. Luedtke, C. R. Mitchell, J. A. LoMenzo, and M. S. Black. "Effect of Selected Process Parameters on Emission Rates of Volatile Organic Chemicals from Carpet." *American Industrial Hygiene Association Journal* 53 (May 1992): 275–282.

Hodgson, A. T., Joan M. Daisey, and Richard A. Grot. "Sources and Source Strengths of Volatile Organic Compounds in a New Office Building." *Journal of the Air and Waste Management Association* 41 (November 1991): 1461–1468.

Hodgson, A. T., Karina Garbesi, Richard G. Septro, and Joan M. Daisey. "Soil-Gas Contamination and Entry of Volatile Organic Compounds into a Home Near a Landfill." *Journal of the Air and Waste Management Association* 42 (March 1992): 277–283.

Hodgson, A. T., J. D. Wooley, and J. M. Daisey. "Emissions of Volatile Organic Compounds from New Carpets Measured in a Large-Scale Environmental Chamber." *Journal of the Air and Waste Management Association* 43 (March 1993): 316–324.

Lawryk, N. J., and C. P. Weisel. "Concentrations of Volatile Organic Compounds in the Passenger Compartments of Automobiles." *Environmental Science and Technology* 30 (March 1996): 810–816.

Little, John C. "Applying the Two-Resistance Theory to Containment Volatilization in Showers." *Environmental Science and Technology* 26 (July 1992): 1341–1349.

Little, John C., Joan M. Daisey, and William W. Nazaroff. "Transport of Subsurface Contaminants into Buildings: An Exposure Pathway for Volatile Organics." *Environmental Science and Technology* 26 (November 1992): 2058–2066.

Otson, Rein, Philip Fellin, and Quang Tran. "VOC's (Volatile Organic Compounds) in Representative Canadian Residences." *Atmospheric Environment* (England) 28, no. 22 (December 1994): 3563–3569.

Pierce, W. M., J. N. Janczewski, Brian Roethlisberger, Mike Pelton, and Kristen Kunstal. "Effectiveness of Auxiliary Air Cleaners in Reducing ETS Components in Offices." *ASHRAE Journal* 38 (November 1996): 51–52+.

Schaeffer, V. H., Bharat-Bhooshan, Shing-Bong Chen, Jay S. Sonenthal, and Alfred T. Hodgson. "Characterization of Volatile Organic Chemical Emissions from Carpet Cushions." *Journal of the Air and Waste Management Association* 46 (September 1996): 813–820.

Sollinger, S., K. Levsen, and G. Wiensch. "Indoor Air Pollution by Organic Emissions from Textile Floor Coverings: Climate Chamber Studies Under Dynamic Conditions." *Atmospheric Environment, Part B* 27B (June 1993): 183–192.

Sollinger, S., et al. "Indoor Air Pollution by Organic Emissions from Textile Floor Coverings: Climate Test Chamber Studies

Under Static Conditions." *Atmospheric Environment,* (England) 28 (August 1994): 2369–2978.

Sullivan, T. F. "EPA Reduces Chemical Emissions from Carpets." *Pollution Engineering* 22 (August 1990): 41–42.

Tancrède, M., Y. Yanagisawa, and R. Wilson. "Volatilization of Volatile Organic Compounds from Showers." *Atmospheric Environment, Part A* 26A, no. 6 (April 1992): 1103–1111.

Tepper, J. S., V. C. Moser, D. L. Costa, M. A. Mason, Nancy Rocke, Zhiski Guo, and R. S. Dyer. "Toxicological and Chemical Evaluation of Emissions from Carpet Samples." *American Industrial Hygiene Association Journal* 56 (February 1995): 158–170.

Trechsel, H. R. "Moisture Control in Buildings." *ASTM Standardization News* 21 (December 1993): 44–49.

Weschler, C. J., A. T. Hodgson, and J. D. Wooley. "Indoor Chemistry: Ozone, Volatile Organic Compounds, and Carpets." *Environmental Science and Technology* 26 (December 1992): 2371–2377.

Health Risks

Lewis, R. G., and L.A. Wallace. "Toxic Organic Vapors in Indoor Air." *ASTM Standardization News* 16 (December 1988): 40–44.

McKone, T. E. "Human Exposure to Volatile Organic Compounds in Household Tap Water: The Indoor Inhalation Pathway." *Environmental Science and Technology* 21 (December 1987): 1194–1201. Discussion, 26 (April 1992): 836–838.

Molhave, Lars, B. Bach, and O. F. Pedersen. "Human Reactions to Low Concentrations of Volatile Organic Compounds." *Environment International* 8 (1987): 117–127.

Molhave, Lars, John G. Jensen, and Soren Larsen. "Subjective Reactions to Volatile Organic Compounds as Air Pollutants." *Atmospheric Environment, Part A* 25A, no. 7 (1991): 1283–1293.

Muller, W. J., and V. H. Schaeffer. "A Strategy for the Evaluation of Sensory and Pulmonary Irritation Due to Chemical Emissions from Indoor Sources." *Journal of the Air and Waste Management Association* 46 (September 1996): 808–812.

Shah, J. J., and H. B. Singh. "Distribution of Volatile Organic Chemicals in Outdoor and Indoor Air." *Environmental Science and Technology* 22 (December 1988): 1381–1388.

Sterling, D. A. "Volatile Organic Compounds in Indoor Air: An Overview of Sources, Concentrations, and Human Effects." In *Indoor Air and Human Health.* Ed. Gammage, R. B., S. B. Kaye, and V. A. Jacobs. Chelsea, MI: Lewis Publishers, 1985, pp. 387–402.

Wallace, L.A., E. D. Pellizzari, T. D. Hartwell, Vicki Davis, L. C. Michael, and Roy W. Whitmore. "The Influence of Personal Activities on Exposure to Volatile Organic Compounds." *Environmental Research* 50 (October 1989): 37–55.

Weschler, C. J., H. C. Shields, and D. Rainer. "Concentrations of Volatile Organic Compounds at a Building with Health and Comfort Complaints." *American Industrial Hygiene Association Journal* 51 (May 1990): 261–268.

Formaldehyde

Adams, L. "EPA Study: Formaldehyde Emissions Less Than Expected." *Wood and Wood Products* 101 (June 1996): 12.

Beck, S. W., and T. H. Stock. "An Evaluation of the Effect of Source and Concentration on Three Methods for the Measurement of Formaldehyde in Indoor Air." *American Industrial Hygiene Association Journal* 51 (January 1990): 14–22.

Broder, Irvin, Paul Corey, Philip Cole, Michael Lipa, Sheldon Mintz, and James R. Nethercott. "Comparison of Health of Occupants and Characteristics of Houses among Control Homes and Homes Insulated with Area Formaldehyde Foam." *Environmental Research* 45 (April 1988): 141–203.

Godish, T. "Residential Formaldehyde Increased Exposure Levels Aggravate Adverse Health Effects." *Journal of Environmental Health* 53 (November-December 1990): 34–37.

Godish, T., and J. Rouch. "Mitigation of Residential Formaldehyde Contamination by Indoor Climate Control." *American Industrial Hygiene Association Journal* 47 (December 1986): 792–797.

Groah, W. J., John Bradfield, Gary Gramp, Rob Rudzinski, and Gary Heroux. "Comparative Response of Reconstituted Wood Products to European and North American Test Methods for Determining Formaldehyde Emissions." *Environmental Science and Technology* 25 (January 1991): 117–122.

Grosjean, D., and E. L. Williams. "A Passive Sampler for Airborne Formaldehyde." *Atmospheric Environment, Part A* 26A, no. 16 (November 1992): 2923–2928.

Hernandez, O., et al. "Risk Assessment of Formaldehyde." *Journal of Hazardous Materials* 39 (November 1994): 161–172.

Ho, M. H., and R. A. Richards. "Enzymatic Method for Determination of Formaldehyde." *Environmental Science and Technology* 24 (February 1990): 201–204.

Hornung, R. W., R. F. Herrick, P. A. Stewart, D. F. Utterback, C. E. Feigley, D. K. Wall, D. E. Douthit, and R. B. Hayes. "An Experimental Design Approach to Retrospective Exposure Assessment." *American Industrial Hygiene Association Journal* 57 (March 1996): 251–256.

Krzyzanowski, M., James Quackenboss, and Michael D. Lebowitz. "Chronic Respiratory Effects of Indoor Formaldehyde Exposure." *Environmental Research* 52 (August 1990): 117–125.

Kulle, T. J., L. R. Sauder, J. R. Hebel, D. J. Green, and M. D. Chatham. "Formaldehyde Dose-Response in Healthy Nonsmokers." *JAPCA* 37 (August 1987): 919–924.

Matthews, T. G., K. W. Fung, B. J. Tromberg, and A. R. Hawthorne. "Impact of Indoor Environmental Parameters on Formaldehyde Concentrations in Unoccupied Research Houses." *Journal of the Air Pollution Control Association* 36 (November 1986): 1244–1249.

Matthews, T. G., A. R. Hawthorne, and C. V. Thompson. "Formaldehyde Sorption and Desorption Characteristics of Gypsum Wallboard." *Environmental Science and Technology* 21 (July 1987): 629–634.

Obee, T. N., and R. T. Brown. "TiO_2 Photocatalysis for Indoor Air Applications: Effects of Humidity and Trace Contaminant Levels on the Oxidation Rates of Formaldehyde, Toluene, and 1.3 Butadiene." *Environmental Science and Technology* 29 (May 1995): 1223–1231.

Passman, F. J. "Formaldehyde Risk in Perspective: A Toxicological Comparison of Twelve Biocides." *Lubrication Engineering* 52 (January 1996): 69–80.

Sexton, Ken, Myrto Petreas, and Kai-Shen Liu. "Formaldehyde Exposures Inside Mobile Homes." *Environmental Science and Technology* 23 (August 1989): 985–988.

Silberstein, Samuel, Richard A. Grot, Kunimichi Ishiguro, and Joseph L. Mulligan. "Validation of Models for Predicting Formaldehyde Concentrations in Residences Due to Pressed-Wood Products." *JAPCA* 38 (November 1988): 1403–1411.

Smulski, S. "Formaldehyde Indoors: Use Reconstituted Wood Products with Lower Emissions." *Progressive Builder* 12 (April 1987): 9–11.

Stock, T. H. "Formaldehyde Concentrations Inside Conventional Housing." *JAPCA* 37 (August 1987): 913–918.

Pesticides

Anderson, D. J., and R. A. Hites. "Chlorinated Pesticides in Indoor Air." *Environmental Science and Technology* 22 (June 1988): 717–720.

———. "Indoor Air: Spatial Variations of Chlorinated Pesticides." *Atmospheric Environment* 23, no. 9 (1989): 2063–2066.

Bukowski, J. A., and L. W. Meyer. "Simulated Air Levels of Volatile Organic Compounds Following Different Methods of Indoor Insecticide Application." *Environmental Science and Technology* 29 (March 1995): 673–676.

Bukowski, J. A., Mark G. Robson, Brian T. Buckley, Daniel W. Russell, and Leroy W. Meyer. "Air Levels of Volatile Organic Compounds Following Indoor Application of an Emulsifiable Concentrate Insecticide." *Environmental Science and Technology* 30 (August 1996): 2543–2546.

Chen, S. Y. S., H. C. Yu, and David D. W. Hwang. "Ventilation Analysis for a VAV System." *Heating/Piping/Air Conditioning* 64 (April 1992): 36–41.

El-Shobokshy, M. S., and F. M. Hussein. "Correlation between Indoor-Outdoor Inhalable Particulate Concentrations and Meteorological Variables." *Atmospheric Environment* 22, no. 12 (1988): 2667–2675.

Nishioka, M. G., H. M. Burkholder, M. S. Brinkman, S. M. Gordon, and R. G. Lewis. "Measuring Transport of Lawn-Applied Herbicide Acids from Turf to Home: Correlation of Dislodgeable 2, 4-D Turf Residues with Carpet Dust, and Carpet Surface Residues." *Environmental Science and Technology* 30 (November 1996): 3313–3320.

Stone, Janis, Steven Padgitt, Wendy Winterstein, Mark Shelley, and Sandra Chisholm. "Iowa Greenhouse Applicators' Perception and Use of Personal Protective Equipment." *Journal of Environmental Health* 57 (October 1994): 16–22.

U.S. Environmental Protection Agency. *A Consumer's Guide to Safer Pesticide Use.* OPA 87-013. Washington, DC: EPA, Office of Public Affairs, 1987.

———. *Assessment of Health Risks to Garment Workers and Certain Home Residents from Exposure to Formaldehyde.* Washington, DC: EPA, Office of Pesticides and Toxic Substances, 1987.

———. *Nonoccupational Pesticide Exposure Study NO PES.* EPA/600/3-90/003. Research Triangle Park, NC: EPA, Atmospheric Research and Exposure Assessment Laboratory, 1990.

Wallace, J. C., L. P. Brzuzy, S. L. Simonich, S. M. Visscher, and R. A. Hites. "Case Study of Organchlorine Pesticides in the Indoor Air of a Home." *Environmental Science and Technology* 30 (September 1996): 2715–2718.

Williams, David T., Cathy Shewchuck, Gary L. Lebel, and Nancy Muir. "Diazinon Levels in Indoor Air After Application for Insect Control." *American Industrial Hygiene Association Journal* 48 (September 1987): 780–785.

Xue, S. "Health Effects of Pesticides: A Review of Epidemiologic Research from the Perspective of Developing Nations." *American Journal of Industrial Medicine* 12 (1987): 269–279.

Noise Pollution

General

Bennett, J. S., et al. "Estimating the Contribution of Individual Work Tasks to Room Concentration: Method Applied to Embalming." *American Industrial Hygiene Association Journal* 57 (July 1996): 599–609.

Blazier, W. E. Jr. "Sound Quality Considerations in Rating Noise from Heating, Ventilating, and Air-Conditioning (HVAC) Systems in Buildings." *Noise Control Engineering Journal* 43 (May-June 1995): 53–63.

Bradley, J. S. "Annoyance Caused by Constant-Amplitude and Amplitude-Modulated Sounds Containing Rumble." *Noise Control Engineering Journal* 42 (November-December 1994): 203–208.

Guenther, F. "IAQ and Noise Control Working Together." *Heating/Piping/Air Conditioning* 68 (January 1996): 59–61+.

Kuhner, Dietrich, Peter Ploner, and Karl R. Bleimann. "Noise Abatement for Electric Arc Furnaces." *Iron and Steel Engineer* 73 (April 1996): 83–86.

Lusk, S. L., D. L. Ronis, and M. J. Kerr. "Predictors of Hearing Protection Use among Workers: Implications for Training Programs." *Human Factors* 37 (September 1995): 635–640.

Schomer, P. D. "25 Years of Progress in Noise Standardization." *Noise Control Engineering Journal* 44 (May-June 1996): 141–148.

Suter, A. H., and D. L. Johnson. "Progress in Controlling Occupational Noise Exposure." *Noise Control Engineering Journal* 44 (May-June 1996): 109–119.

Sick Building Syndrome

Building Design

Baillie, Sheila. "High-Tech or Healthful Housing: Must We Make a Choice?" *National Forum* 76 (Spring 1996): 38–40+.

Cox, B. G., D. T. Mage, and F. W. Immerman. "Sample Design Considerations for Indoor Air Exposure Surveys." *JAPCA* 38 (October 1988): 1266–1270.

Dubin, F. S. "Integrated Building Systems: Adding the Human Element." *Consulting-Specifying Engineer* 8 (July 1990): 57–62+.

Gill, K. E. "HVAC Design for Isolation Rooms." *Heating/Piping/Air Conditioning* 66 (February 1994): 5–8+.

———. "IAQ and Air Handling Unit Design." *Heating/Piping/Air Conditioning* 68 (January 1996): 49–54. Correction, 68 (April 1996): 42.

Hayter, A. J., and M. M. Dowling. "Experimental Designs and Admission Rate Modeling for Chamber Experiments. *Atmospheric Environment, Part A* 27A, no. 14 (October 1993): 2225–2234.

Kennedy, C. R., and A. Distefano. "Improving Building IAQ Reduces HVAC Energy Costs." *Heating/Piping/Air Conditioning* 63 (November 1991): 75–79.

Ruppersberger, J. S. "Concrete Blocks' Adverse Effects on Indoor Air and Recommended Solutions." *Journal of Environmental Engineering* 121 (April 1995): 348–356.

Schaeffer, David J., Roger A. Deem, and Edward W. Novak. "Indoor Firing Range Air Quality: Results of a Facility Design Survey." *American Industrial Hygiene Association Journal* 51 (February 1990): 84–87.

Scofield, C. M., and N. H. Des Champs. "HVAC Design for Classrooms: Divide and Conquer." *Heating/Piping/Air Conditioning* 65 (May 1993): 53–59.

Sick Building Syndrome—General

Anderson, R. C. "Indoors, the Newest Polluted Space." *Pollution Engineering* 24 (April 1, 1992): 58–60.

Bachmann, Max O., and Jonathan Myers. "Influences on Sick Building Syndrome Symptoms in Three Buildings." *Social Science and Medicine* 40, no. 2 (1995): 245–251.

Bain, Peter, and Chris Baldry. "Sickness and Control in the Office: The Sick Building Syndrome." *New Technology, Work, and Employment* 10 (March 1995): 19–31.

Bisio, A. "Sick Building Syndrome." *Chemical Engineering* (England), no. 474 (May 24, 1980): 18–21.

"Combatting 'Sick Building Syndrome.'" *Civil Engineering* 64 (January 1994): 20–21.

Daisy, J. M., A. T. Hodgson, W. J. Fisk, M. J. Mendell, and J. T. Brinks. "Volatile Organic Compounds in Twelve California Office Buildings: Classes, Concentrations, and Sources." *Atmospheric Environment* (England) 28, no. 22 (December 1994): 3557–3562.

DeKoster, J. A., and P. S. Thorne. "Bioaerosol Concentrations in Noncompliant, Compliant, and Intervention Homes in the Midwest." *American Industrial Hygiene Association Journal* 56 (June 1995): 573–580.

Ekberg, L. E. "Volatile Organic Compounds in Office Buildings." *Atmospheric Environment* (England) 28, no. 22 (December 1994): 3571–3575.

"Emission from Mold/Fungus May Be Culprits in IAQ." *Journal of Envirionmental Health* 38 (October 1995): 32.

Gardner, T. F. "Is the Ventilation Engineer Responsible for Sick Building Syndrome Damages?" *ASHRAE Journal* 32 (August 1990): 22–25. Discussion, 33 (January 1991): 13.

Giampetro-Meyer, Andrea. "Rethinking Workplace Safety: An Integration and Evaluation of Sick Building Syndrome and Fetal Protection Cases." *UCLA Journal of Environmental Law and Policy* 8, no. 1 (1988): 1–29.

Guirguis, S., G. Rajhans, D. Leong, and L. Wong. "A Simplified IAQ Questionnaire to Obtain Useful Data for Investigating Sick Building Complaints." *American Industrial Hygiene Association Journal* 52 (August 1991): A434-A437.

Hansen, W. "Indoor Air Quality and Energy Conservation." *Energy Engineering* 89, no. 6 (1992): 16–31.

Hays, S. M., and N. Ganick. "How to Attack IAQ Problems." *Heating/Piping/Air Conditioning* 64 (April 1992): 43–47+.

Hodgson, M. J., and C. A. Hess. "Doctors, Lawyers, and Building Associated Diseases." *ASHRAE Journal* 34 (February 1992): 25–32.

Kim, I. "Sick Building Syndrome: MEs Seek a Cure." *Mechanical Engineering* 112 (November 1990): 53–56. Discussion, 113 (January 1991): 6.

Kjaergaard, Soren, Lars Molhave, and Ole F. Pedersen. "Human Reactions to a Mixture of Indoor Air Volatile Organic Compounds." *Atmospheric Environment, Part A* 25A, no. 8 (1991): 1417–1426.

Kreiss, K. "The Epidemiology of Building Related Complaints and Illness." *Occupational Medicine State of the Art Reviews* 4, no. 4 (1989): 575–592.

McCarthy, S. M. "An Overview of Sick Building Investigations and Legal Issues." *Energy Engineering* 90, no. 6 (1993): 43–58.

Meckler, M. "Employing ASHRAE Standard 62-1989 in Urban Building Environments." *Energy Engineering* 88, no. 3 (1991): 34–60.

Mikatavage, M. A., V. E. Rose, Ellen Funkhouser, R. K. Oestenstad, Kenneth Dillon, and K. D. Reynolds. "Beyond Air Quality—Factors That Affect Prevalence Estimates of Sick Building Syndrome." *American Industrial Hygiene Association Journal* 56 (November 1995): 1141–1146.

Newman, A. "Do Microbes Contribute to Sick Building Syndrome?" *Environmental Science and Technology* 29 (October 1995): 450A.

Pollina, Ronald R. "Sick Building Syndrome: Sleeping Giant of the '90s." *Site Selection* 38 (April 1993): 442–445.

Rand, Ted G. "Sick Building Syndrome Needs Airing." *Insurance Review* 51 (May 1990): 33–34+.

Rothweiler, Heinz, Patrick A. Wagner, and Christian Schlatter. "Comparison of Tenax TA and Carbotrap for Sampling and Analysis of Volatile Organic Compounds in Air." *Atmospheric Environment, Part B* 25B, no. 2 (1991): 231–235.

Ryan, T. J. "Contrast Specs for Indoor Air Quality Investigations." *Professional Safety* 38 (July 1993): 20–23.

Scarry, R. L. "Looking into Sick Buildings." *Heating/Piping/Air Conditioning* 66 (July 1994): 63–65.

Shearer, R. W. "Building-Associated Illness." *Professional Safety* 36 (December 1991): 15–21.

"Sick Building Syndrome." *InTech* 42 (January 1995): 78–79.

"Sick Buildings: What We Know, What We Don't." *ASHRAE Journal* 37 (December 1995): 10+.

Smith, S. C. "Controlling 'Sick Building Syndrome.'" *Journal of Environmental Health* 53 (November-December 1990): 22–23.

"Study Suggests Some VOCs Caused by Molds and Fungi." *ASHRAE Journal* 37 (October 1995): 12.

Sundell, J. "What We Know and Don't Know about Sick Building Syndrome." *ASHRAE Journal* 38 (June 1996): 51–54+.

Totty, Patrick. "When the Office Is Sick: How Indoor Pollutants Became the Scourge of the '80s." *San Francisco Business* 26 (December 1989): 14–19.

Wolkoff, P., and P. A. Nielson. "A New Approach for Indoor Climate Labeling of Building Materials—Emission Testing, Modeling, and Comfort Evaluation." *Atmospheric Environment* (England) 30, no. 15 (August 1996): 2679–2689.

Control and Management

Batterman, Stuart, and Harriet Burge. *HVAC Systems as Emission Sources Affecting Indoor Air Quality: A Critical Review: A Project Summary.* 95-0594-M. Research Triangle Park, NC: Environmental Protection Agency, Air and Energy Engineering Research Laboratory, 1995, microfiche.

Bennett, J. S., et al. "Estimating the Contribution of Individual Work Tasks to Room Concentration: Method Applied to Embalming." *American Industrial Hygiene Association Journal* 57 (July 1996): 599–609.

Cheremisinoff, P. N., L. M. Ferrante, R. P. Ouellette, and N. J. Van Houten. "Managing In-Plant Environmental Problems." *Pollution Engineering* 23 (April 1991): 52–58.

Estill, C. F., and A. B. Spender. "Case Study: Control of Methylene Chloride Exposures During Furniture Stripping." *American Industrial Hygiene Association Journal* 57 (January 1996): 43–49.

"Filtration Brings IAQ/Humidity Control Cost Savings to School." *Heating/Piping/Air Conditioning* 68 (August 1996): 23–25+.

Franke, J. E., and R. A. Wadden. "Indoor Contaminant Emission Rate Characterized by Source Activity Factors." *Environmental Science and Technology* 21 (January 1987): 45–51.

Halm, P. E. "Controlling Indoor Air Quality." *Consulting-Specifying Engineer* 5 (January 1989): 42–49.

"IAQ Control by Source Elimination or Modification." *ASHRAE Journal* 29 (July 1987): 23–28.

"IAQ to Emphasize Practical Approaches (Atlanta, GA, October 6–8, 1996)." *ASHRAE Journal* 38 (August 1996): 27.

Jensen, P. A., W. F. Todd, M. E. Hart, R. L. Mickelsen, and D. M. O'Brien. "Evaluation and Control of Worker Exposure to Fungi in a Beet Sugar Refinery." *American Industrial Hygiene Association Journal* 54 (December 1993): 742–748.

Katzel, J. "Achieving Indoor Air Quality through Contaminant Control." *Plant Engineering* 49 (July 10, 1995): 42–46.

Lees-Haley, P. R. "Indoor Air Quality: Recommendations to Management and Owners." *Professional Safety* 38 (October 1993): 33–35.

McKone, T. E., and K. T. Bogen. "Predicting the Uncertainties in Risk Assessment." *Environmental Science and Technology* 25 (October 1991): 1674–1681.

McNall, P. E. "The HVAC Engineer and Indoor Air Quality." *Heating/Piping/Air Conditioning* 60 (February 1988): 65–70.

Naumann, B. D., E. V. Sargent, B. S. Starkman, W. J. Fraser, G. T. Becker, and G. D. Kirk. "Performance-Based Exposure Control

Limits for Pharmaceutical Active Ingredients." *American Industrial Hygiene Association Journal* 57 (January 1996): 33–42.

Nero, A. V. "Controlling Indoor Air Pollution." *Scientific American* 258 (May 1988): 42–48.

Sexton, K., and S. B. Hayward. "Source Apportionment of Indoor Air Pollution." *Atmospheric Environment* 21, no. 2 (1987): 407–418.

Thorsen, M. A., and L. Molhave. "Elements of a Standard Protocol for Measurements in the Indoor Atmospheric Environment." *Atmospheric Environment* 21, no. 6 (1987): 1411–1416.

Trichenov, B. A., L.A. Sparks, J. B. White, and M. D. Jackson. "Evaluating Sources of Indoor Air Pollution." *Journal of the Air and Waste Management Association* 40 (April 1990): 487–492.

Van Wormer, M. B. "Use Air Quality Auditing as an Environmental Management Tool." *Chemical Engineering Progress* 87 (November 1991): 62–67.

Williams, D. T., Cathy Shewchuck, Guy L. Lebel, and Nancy Muir. "Diazinon Levels in Indoor Air After Periodic Application for Insect Control." *American Industrial Hygiene Association Journal* 48 (September 1987): 780–785.

Measurement and Analysis

Baughman, K. W., and D. H. Love. "Industrial Hygiene Chemistry: The Method Development Approach to Air Analysis." *Analytical Chemistry* 65 (May 15, 1993): 480A–484A+.

Burroughs, G. E., and W. J. Woodfin. "On-Site Screening for Benzene in Complex Environments." *American Industrial Hygiene Association Journal* 56 (September 1995): 874–882.

Cohen, M. A., P. B. Ryan, Yukio Yanagisawa, J. D. Spengler, Halick Özkaynak, and P. S. Epstein. "Indoor/Outdoor Measurements of Volatile Organic Compounds in the Kanawha Valley of West Virginia." *JAPCA* 39 (August 1989): 1086–1093.

Contant, C. F., T. H. Stock, P. A. Buffler, A. H. Holguin, B. M. Gehan, and D. J. Kotchmar. "The Estimation of Personal Exposures to Air Pollutants for a Community-Based Study of Health Effects in Asthmatics—Exposure Model." *JAPCA* 37 (May 1987): 587–594.

Draxler, R. R. "Estimating Emissions from Air Concentration Measurements." *JAPCA* 37 (June 1987): 708–714.

Fischer, D., and C. G. Uchrin. "Laboratory Simulation of VOC Entry into Residence Basements from Soil Gas." *Environmental Science and Technology* 30 (August 1996): 2598–2603.

Grenier, M. G., Stephen G. Hardcastle, Gopal Kunchur, and Kevin Butler. "The Use of Tracer Gases to Determine Dust Dispersion Patterns and Ventilation Parameters." *American Industrial Hygiene Association Journal* 53 (June 1992): 387–394.

Jayanty, R. K. M., and B. W. Gray Jr. "Summary of the 1993 EPA/AandWMA International Symposium: Measurement of Toxic and Related Air Pollutants [Durham, NC, May 3–7, 1993]." *Journal of the Air and Waste Management Association* 44 (March 1994): 254–259.

Jones, J., et al. "Performance Analysis for Commercially Available Sensors." *Journal of Architectural Engineering* 3 (March 1997): 25–31.

Lindstrom, A. B., and J. D. Pleil. "A Methodological Approach for Exposure Assessment Studies in Residences Using Organic Compound-Contaminated Water." *Journal of the Air and Waste Management Association* 46 (November 1996): 1058–1066.

Lioy, P. J., J. M. Waldman, T. Buckley, J. Butler, and C. Pietarinen. "The Personal Indoor and Outdoor Concentrations of PM-10 Measured in an Industrial Community During the Winter." *Atmospheric Environment, Part B* 24B, no. 1 (1990): 57–66.

Malachowski, M. S., S. P. Levine, G. Herrin, R. C. Spear, M. Yost, and Z. Yi. "Workplace and Environmental Air Contaminant Concentrations Measured by Open Path Fourier Transform Infrared Spectroscopy: A Statistical Process Control Technique to Detect Changes from Normal Operating Conditions." *Journal of the Air and Waste Management Association* 44 (May 1994): 673–682.

Matney, M. L., et al. "Pyrolysis-Gas Chromatography/Mass Spectrometry Analyses of Biological Particles Collected During Recent Space Shuttle Missions." *Analytical Chemistry* 66 (September 15, 1994): 2820–2828.

Moschandreas, D. J., and D. S. O'Dea. "Measurement of Perchloroethylene Indoor Air Levels Caused by Fugitive Emissions from Unvented Dry-to-Dry Dry Cleaning Units." *Journal of the Air and Waste Management Association* 45 (February 1995): 111–115.

Mumford, J. L., D. B. Harris, and Katherine Williams. "Indoor Air Samplings and Mutagenicity Studies of Emissions from Unvented Coal Combustion." *Environmental Science and Technology* 21 (March 1987): 308–311.

"The New Standard Environmental Inventory Questionnaire for Estimation of Indoor Constructions." *JAPCA* 39 (November 1989): 1411–1419.

Nicas, M. "Estimating Exposure Intensity in an Imperfectly Mixed Room." *American Industrial Hygiene Association Journal* 57 (June 1996): 542–550.

Olcerst, R. "Measurement of Outdoor and Recirculated Air Percentages by Carbon Dioxide Tracer." *American Industrial Hygiene Association Journal* 55 (June 1994): 525–528.

Riggin, R. M., and B. A. Petersen. "Sampling and Analysis Methodology for Semivolatile and Nonvolatile Organic Compounds in Air." In *Indoor Air and Human Health.* Ed. Gammage, R. B., S. B. Kaye, and V. A. Jacobs. Chelsea, MI: Lewis Publishers, 1985, pp. 351–358.

Rosebrook, D. D., and G. Worm. "Personal Exposures, Indoor-Outdoor Relationships, and Breath Levels of Toxic Air Pollutants Measured for 355 Persons in New Jersey (Discussion of Article by L. A. Wallace, et al.)" *Atmospheric Environment, Part A* 27A, no. 14 (October 1993): 2243–2249.

Schroll, C. "Monitoring Hazardous Atmospheres." *Fire Engineering* 150 (February 1997): 74–79.

Sinclair, J. D., L. A. Psota-Kelty, G. A. Peins, and A. O. Ibidunni. "Indoor/Outdoor Relationships of Airborne Ionic Substances: Comparison of Electronic Equipment Room and Factory Environments." *Atmospheric Environment, Part A* 26A, no. 5 (April 1992): 871–882.

Thompson, C. V., R. A. Jenkins, and C. E. Higgins. "A Thermal Desorption Method of the Determination of Nicotine in Indoor Environments." *Environmental Science and Technology* 23 (April 1989): 429–435.

Thorsen, M. A., and L. Molhave. "Elements of a Standard Protocol for Measurements in the Indoor Atmospheric Environment." *Atmospheric Environment* 21, no. 6 (1987): 1411–1416.

Tikiusis, Tony, Michael R. Phibbs, and Kenneth L. Sonnenberg. "Quantitation of Employee Exposure to Emission Products Generated by Commercial-Scale Processing of Polyethylene." *Ameri-*

can Industrial Hygiene Association Journal 56 (August 1995): 809–814.

Todd, L., and G. Ramachandran. "Evaluation of Algorithms for Tomographic Reconstruction of Concentrations in Indoor Air." *American Industrial Hygiene Association Journal* 55 (May 1994): 403–417.

———. "Evaluation of Optical Source-Detector Configurations for Tomographic Reconstruction of Chemical Concentrations in Indoor Air." *American Industrial Hygiene Association Journal* 55 (December 1994): 1133–1143.

Traynor, G. W. "Field Monitoring Design Considerations for Assessing Indoor Exposures to Combustion Pollutants." *Atmospheric Environment* 21, no. 2 (1987): 377–383.

Wallace, J. C., Ilora Basu, and Ronald A. Hites. "Sampling and Analysis Artifacts Caused by Elevated Indoor Air Polychlorinated Biphenyl Concentrations." *Envirionmental Science and Technology* 30 (September 1996): 2730–2734.

Wide, P., et al. "An Air-Quality Sensor System with Fuzzy Classification." *Measurement Science and Technology* 8 (February 1997): 138–146.

Wilson, N. K., M. R. Kuhlman, J. C. Chuang, G. A. Mack, and J. E. Horves Jr. "A Quiet Sampler for the Collection of Semivolatile Organic Pollutants in Indoor Air." *Environmental Science and Technology* 23 (September 1989): 1112–1116.

Yu, R. C., and R. J. Sherwood. "The Relationship between Urinary Elimination, Airborne Concentration, and Radioactive Hand Contamination for Workers Exposed to Uranium." *American Industrial Hygiene Association Journal* 57 (July 1996): 615–620.

Ventilation

Berlin, G. "Achieving Proper Humidity Levels in the Plant." *Plant Engineering* 49 (November 6, 1995): 42–44.

Breum, N. O., and E. Orhede. "Dilution Versus Displacement Ventilation—Environmental Conditions in a Garment Sewing Plant." *American Industrial Hygiene Association Journal* 55 (February 1994): 140–148.

Carlton-Foss, J. A. "Office Ventilation." *ASHRAE Journal* 30 (September 1988): 24–28.

Garrison, R. P., Kiyoung Lee, and Chulhong Park. "Contaminant Reduction by Ventilation in a Confined Space Model-Toxic Concentrations Versus Oxygen Deficiency." *American Industrial Hygiene Association Journal* 52 (December 1991): 542–546.

Hansen, S. J. "Indoor Air Quality: Is Increased Ventilation the Answer?" *Energy Engineering* 86, no. 6 (1989): 33–46.

Jiang, Z., Qingyan Chen, and F. Haghighat. "Airflow and Air Quality in a Large Enclosure." *Journal of Solar Energy Engineering* 117 (May 1995): 114–122.

Lage, J. L., A. Bejan, and R. Anderson. "Removal of Contaminant Generated by a Discrete Source in a Slot Ventilated Enclosure." *International Journal of Heat and Mass Transfer* 35 (May 1992): 1169–1180.

MacCracken, C. D. "Cold Air Systems: Sleeping Giants." *Heating/Piping/Air Conditioning* 66 (April 1994): 59–62.

Matthews, T. G., D. L. Wilson, C. V. Thompson, K. P. Monar, and C. S. Dudney. "Impact of Heating and Air Conditioning System Operation and Leakage on Ventilation and Intercompartment Transport: Studies in Unoccupied Tennessee Valley Homes." *Journal of the Air and Waste Management Association* 40 (February 1990): 194–198.

Moffatt, S. "Backdrafting Woes (Chimney Drafting)." *Progressive Builder* 11 (December 1986): 25–30+.

Schuler, M. "Dual Fan, Dual-Duct System Meets Air Quality, Energy-Efficient Needs." *ASHRAE Journal* 38 (March 1996): 39–41.

Simpson, C. A. "Mist Collection Systems Removes Contaminants." *Pollution Engineering* 25 (October 15, 1993): 15–16.

Sterling, Elia, Chris Collett, and James Ross. "Dilution Ventilation to Accommodate Smoking: A Case Study." *Heating/Piping/Air Conditioning* 68 (November 1996): 81+.

Sun, T.-Y. "Air Balance in a Conditioned Space and Outdoor Air Flow Control." *Heating/Piping/Air Conditioning* 68 (October 1996): 73–75.

Taylor, S. T. "Series Fan-Powered Boxes: Their Impact on Indoor Air Quality and Comfort." *ASHRAE Journal* 38 (July 1996): 44–50.

Warden, D. "Dual Fan, Dual Duct Systems." *ASHRAE Journal* 38 (January 1996): 36–41.

Air Purification

Aaronson, E. L., and F. Fenci. "High-Efficiency Filtration Meets IAQ Goals." *Heating/Piping/Air Conditioning* 66 (December 1994): 69–72.

Bradley, J. S. "Disturbance Caused by Residential Air Conditioning Noise. *Journal of the Acoustical Society of America* 93, pt. 1 (April 1993): 1978–1986.

Brown, William T., Glen Chamberlin, and Jerry M. LaGrou. *Air Cleaning Systems and Indoor Air Quality: A Review.* USA CERL Technical Report, FE-93/10. Springfield, VA: U.S. Army Corps of Engineers, Construction Engineering Research Laboratories, 1995, 1 vol., microfiche.

Davis, W. T., Catherine Cornell, and Maureen Dever. "Comparison of Experimental and Theoretical Efficiencies of Residential Air Filters." *Tappi Journal* 77 (September 1994): 180–186.

"Filtration Brings IAQ Control Cost Savings to School." *Heating/Piping/Air Conditioning* 68 (August 1996): 23–25+.

Goyer, N. "Chemical Contaminants in Office Buildings." *American Industrial Hygiene Association Journal* 51 (December 1990): 615–619.

Imperato, P. J. "Legionellosis and the Indoor Environment." *Bulletin of New York Academy of Medicine* 57, no. 10 (1981): 922–935.

Joffe, M. A. "Chemical Filtration of Indoor Air: An Application Primer." *ASHRAE Journal* 38 (February 1996): 42–44+.

"Killer Bug That Puzzled Scientists: Illness of American Legion Conventioneers in Philadelphia." *U.S. News and World Report* 81 (August 16, 1976): 31–32.

Liu, R.-T., and M. A. Huza. "Filtration and Indoor Air Quality: A Practical Approach." *ASHRAE Journal* 37 (February 1995): 18–23.

Liu, Zeming, Janet Stout, Lawrence Tedesco, Marcie Boldin, Charles Hwang, and Victor Yu. "Efficacy of Ultraviolet Light in Preventing Legionella Colonization of a Hospital Water Distribution Center." *Water Research* 29 (October 1995): 2275–2280.

Mathews, T., et al. "Mystery Fever: Illness of American Legionnaires after Philadelphia Convention." *Newsweek* 88 (August 16, 1976): 16–18+.

Nazaroff, W. "Engineering Solutions to Indoor Air Quality Problems [Special Issue]." *Journal of the Air and Waste Management Association* 46 (September 1996): 805–908.

Pearson, W. J. "Indoor Air Quality: An Increasing Concern for the Paint Industry." *Modern Paint and Coatings* 82 (January 1993): 24–27.

Pierce, W. M., J. N. Janczewski, Brian Roethlisberger, Mike Pelton, and Kristen Kunstel. "Effectiveness of Auxiliary Air Cleaners in Reducing ETS Components in Offices." *ASHRAE Journal* 38 (November 1996): 51–52+.

Turner, F. "Legionella: What to Monitor and How Often." *ASHRAE Journal* 37 (December 1995): 6+.

Van Osdell, D. W., and L. E. Sparks. "Carbon Adsorption for Indoor Air Cleaning." *ASHRAE Journal* 37 (February 1995): 34–40.

Vine, E. "Air-to-Air Heat Exchangers and the Indoor Environment." *Energy* 12 (December 1987): 1209–1215.

Selected Journals

The journals listed below publish articles on many aspects of indoor pollution. Information on indoor pollution is found in a wide variety of journals. For new journals and additional information, please consult *Ulrich's International Periodicals Directory* 1998, 36th ed. (New York: R. R. Bowker, 1997), 5 volumes. Information in the journals listed is arranged as in the following sample entry.

Journal Title

1. Editor
2. Year first published
3. Frequency of publication
4. Code
5. Special features
6. Address of publisher

Air and Waste

1. Harold M. Englund
2. 1951
3. Monthly

4. ISSN 1073-161X
5. Advertising, book reviews, abstracts, bibliographies, charts, illustrations, statistics, index
6. Air and Waste Management Association
   Three Gateway Center, Four West
   Pittsburgh, PA 15222

Air and Water Pollution Control

1. Randy Kebetin
2. 1986
3. Biweekly
4. ISSN 0890-0396
5. —
6. The Bureau of National Affairs, Inc.
   1231 25th Street NW
   Washington, DC 20037

Air Infiltration Review

1. J. Blacknell
2. 1979
3. Quarterly
4. ISSN 0143-6643
5. Bibliographies, book reviews
6. International Energy Agency
   Air Infiltration and Ventilation Centre
   Sovereign Court, Sir William Lyon's Road
   Coventry Warks CV447EZ
   England

American Industrial Hygiene Association Journal

1. Samuel Elkin
2. 1940
3. Monthly
4. ISSN 0002-8894
5. Advertising, book reviews, bibliographies, charts, illustrations, index
6. American Industrial Hygiene Association
   2700 Prosperity Avenue, Suite 250
   Fairfax, VA 22031-4307

American Journal of Public Health

1. Dr. Michael Ibrahim
2. 1911
3. Monthly
4. ISSN 0090-0036
5. Advertising, charts, illustrations, index
6. American Public Health Association
   1015 15th Street NW
   Washington, DC 20005

American Review of Respiratory and Critical Care Medicine

1. Robert A. Klocke
2. 1917
3. Monthly
4. ISSN 1073-449X
5. Advertising, abstracts, book reviews, bibliographies, charts, illustrations, statistics, index, cumulative index
6. American Lung Association
   1740 Broadway
   New York, NY 10019-4374

ASHRAE Journal (Journal of the American Society of Heating, Refrigerating, and Air-Conditioning Engineers)

1. William R. Coker
2. 1914
3. Monthly
4. ISSN 0001-2491
5. Abstracts, bibliographies, book reviews, charts, illustrations, trade literature, index
6. American Society of Heating, Refrigerating, and Air-Conditioning Engineers, Inc.
   1791 Tullie Circle, N.E.
   Atlanta, GA 30329

ASTM Standardization News

1. K. Riley
2. 1973
3. Monthly
4. ISSN 0090-1210
5. Charts, illustrations, index

6. American Society for Testing and Materials
   1916 Race Street
   Philadelphia, PA 19103

Atmospheric Environment

1. Editorial Board
2. 1967
3. 24 issues per year
4. ISSN 1352-2310
5. Advertising, book reviews, charts, illustrations, index
6. Elsevier Science
   660 White Plains Road
   Tarrytown, NY 10591-5153

Business and Health

1. Joe Burns
2. 1983
3. 12 issues per year
4. ISSN 0739-9413
5. Advertising and book reviews
6. Medical Economics Publishing Co., Inc.
   Five Paragon Drive
   Montvale, NJ 07645

Carcinogenesis

1. Editorial Board
2. 1980
3. Monthly
4. ISSN 0143-3334
5. Advertising, book reviews, illustrations, index
6. Oxford University Press
   Oxford Journals
   Walton Street
   Oxford OX2 6DP
   England

Chemistry and Industry

1. A. Miller
2. 1881
3. Semimonthly (twice monthly)

4. ISSN 0009-3068
5. Advertising, book reviews, bibliographies, charts, illustrations, index
6. American Chemical Society
   Member and Subscriber Services
   Box 3337
   Columbus, OH 43210

Clinical and Experimental Allergy

1. A. B. Kay, S. T. Holgate, Allen Stevens, and Sarah Pollard
2. 1971
3. Monthly
4. ISSN 0954-7894
5. Book reviews, illustrations, index
6. Blackwell Science Ltd.
   Osney Mead
   Oxford OX2 OEL
   England

Common Cause Magazine

1. Vicki Kemper
2. 1980
3. Quarterly
4. ISSN 0271-9592
5. —
6. Common Cause
   2030 M Street NW
   Washington, DC 20036

Contamination Control Abstracts

1. R. W. Newbold
2. 1986
3. Quarterly
4. ISSN 0952-1542
5. —
6. Particle Science and Technology Information Service
   Department of Chemical Engineering
   University of Technology

Loughborough, Leics LE11 3TU
England

CQ Researcher

1. Sandra Stencil
2. 1923
3. 4 issues per month
4. ISSN 1056-2036
5. Book reviews, charts, index
6. Congressional Quarterly, Inc.
   1414 22nd Street NW
   Washington, DC 20037

Environmental Forum

1. Stephen Dujack
2. 1981–1986; resumed 1988
3. Bimonthly
4. ISSN 0731-5732
5. —
6. Environmental Law Institute
   1616 P Street SW, Suite 200
   Washington, DC 20036

Environmental Health

1. Claire Brown
2. 1895
3. Monthly
4. ISSN 0013-9270
5. Advertising, book reviews, bibliographies, charts, illustrations, statistics, trade literature, index
6. Chadwick House Group, Ltd.
   Chadwick Court, 15 Hatfields
   London SE1 8DJ
   England

Environmental Health Perspective

1. George W. Lucier and Gary E. R. Hook
2. 1993
3. Monthly
4. —
5. —

6. U.S. Department of Health and Human Services
   National Institute of Environmental Health Sciences
   Box 12233
   Research Triangle Park, NC 27709

Environmental Progress

1. Gary F. Bennett
2. 1982
3. Quarterly
4. ISSN 0278-4491
5. —
6. American Institute of Chemical Engineers
   345 E. 47th Street
   New York, NY 10017

Environmental Research

1. Philip J. Landrigan
2. 1967
3. 8 issues per year
4. ISSN 0013-9351
5. Advertising, book reviews, illustrations, index
6. Academic Press, Inc.
   Journal Division
   525 B Street, Suite 1900
   San Diego, CA 92101-4495

Environmental Science and Technology

1. William H. Glaze
2. 1967
3. Monthly
4. ISSN 0013-936X
5. Advertising, book reviews, abstracts, bibliographies, charts, illustrations, statistics, trade literature, index
6. American Chemical Society
   1155 16th Street NW
   Washington, DC 20036

EPA Journal

1. John Heritage
2. 1975
3. Bimonthly

4. ISSN 0145-1189
5. Advertising
6. U.S. Environmental Protection Agency
   Office of Public Affairs, Waterside Mall
   401 M Street SW
   Washington, DC 20460

Harvard Environmental Law Review

1. —
2. 1976
3. 2 issues per year
4. ISSN 0147-8257
5. —
6. Harvard University Law School
   Publications Center
   Hastings Hall
   Cambridge, MA 02138

Human Ecology Forum

1. Judy Stewart
2. 1970
3. Quarterly
4. ISSN 0018-7178
5. Bibliographies, charts, illustrations, index
6. New York State College of Human Ecology
   1150 Comstock Hall
   Cornell University
   Ithaca, NY 14850-0998

JAPCA (Journal of the Air Pollution Control Association)

1. Randy Kubetin
2. 1980
3. Biweekly
4. ISSN 0196-7150
5. Index
6. The Bureau of National Affairs, Inc.
   1231 25th Street NW
   Washington, DC 20037

Journal of Air Pollution Control

1. Randy Kubetin
2. 1980
3. Biweekly
4. ISSN 0196-7150
5. Index
6. The Bureau of National Affairs, Inc.
   1231 25th Street NW
   Washington, DC 20037

Journal of Allergy and Clinical Immunology

1. Dr. Philip S. Norman
2. 1929
3. Monthly
4. ISSN 0091-6749
5. Advertising, charts, illustrations, index
6. Mosby-Year Book, Inc.
   11830 Westline Industrial Drive
   St. Louis, MO 63146-3318

Journal of Environmental Health

1. Simonne Gallaty
2. 1938
3. 10 issues per year
4. ISSN 0022-0892
5. Book reviews, charts, illustrations, index
6. National Environmental Health Association
   720 S. Colorado Boulevard, N. 970 S. Tower
   Denver, CO 80222-1904

Journal of Epidemiology and Community Health

1. S. Donnan
2. 1947
3. Bimonthly
4. ISSN 0143-005X
5. Book reviews, bibliographies, charts, illustrations, index
6. BMJ Publishing Group
   Box 480
   Franklin, MA 02038

Journal of Hazardous Materials

1. G. F. Bennett and R. E. Britter
2. 1976
3. 15 issues per year
4. ISSN 0304-3894
5. Advertising, book reviews
6. Elsevier Science
   Regional Sales Office
   Box 945
   New York, NY 10159-0945

Journal of the American Medical Association

1. George D. Lundberg
2. 1848
3. Weekly
4. ISSN 0098-7484
5. Book reviews, bibliographies, charts, illustrations, index (twice annually)
6. American Medical Association
   515 N. State Street
   Chicago, IL 60610

Journal of the Association of Analytical Chemistry

1. Royce W. Murray
2. 1929
3. Twice monthly
4. ISSN 0003-2700
5. Advertising, book reviews, bibliographies, charts, illustrations, trade literature, index
6. American Chemical Society
   1155 16th Street NW
   Washington, DC 20036

Mycopathologia

1. Editorial Board
2. 1938
3. Monthly
4. ISSN 0301-486X
5. Advertising, book reviews, illustrations
6. Kluwer Academic Publishers
   Box 358

Accord Station
Hingham, MA 02018-0358

New England Journal of Medicine

1. Jerome Kassirer
2. 1812
3. Weekly
4. ISSN 0028-4793
5. Book review, bibliographies, charts, illustrations, statistics, index (twice annually)
6. Massachusetts Medical Society
   Publishing Division
   10 Shattuck Street
   Boston, MA 02115

New Scientist

1. Alun Anderson
2. 1956
3. Weekly
4. ISSN 0262-4079
5. Advertising, book reviews, charts, illustrations, patents, quarterly index
6. King's Reach Tower
   Stamford Street
   London SE1 9LS
   England

New York Academy of Medicine Bulletin

1. Dr. Robert Haggarty
2. 1925
3. Twice yearly
4. ISSN 0028-7091
5. Book reviews, charts, illustrations, index
6. New York Academy of Medicine
   2 E. 103rd Street
   New York, NY 10029

New York Academy of Sciences Annals

1. Bill Boland
2. 1823
3. Irregular

4. ISSN 0077-8923
5. Bibliographies, charts, illustrations, index, cumulative index
6. New York Academy of Sciences
   2 E. 63rd Street
   New York, NY 10021

Nuclear Safety

1. E. G. Silver
2. 1959
3. Biannually
4. ISSN 0029-5604
5. Book reviews, charts, illustrations, index
6. U.S. Department of Energy
   Office of Scientific and Technical Information
   Box 62
   Oak Ridge, TN 37831

Nuclear Technology

1. William Vogelsang
2. 1965
3. Monthly
4. ISSN 0029-5450
5. Book reviews, charts, illustrations, statistics, index
6. American Nuclear Society
   555 N. Kensington Avenue
   La Grange Park, IL 60525

Occupational Health and Safety

1. Mark Hartley
2. 1976
3. Monthly
4. ISSN 0362-4064
5. Advertising, book reviews, charts
6. Stevens Publishing Corporation
   3630 J. H. Kultgen Freeway
   Waco, TX 76706

Oncology

1. P. P. Carbone
2. 1948

3. Bimonthly
4. ISSN 0030-2414
5. Advertising, book reviews, bibliographies, charts, illustrations, index
6. S. Karger AG
   Allschwilerstrasse 10
   P.O. Box CH 4009
   Basel, Switzerland

Pollution Engineering

1. Diane Pirocanac
2. 1969
3. 13 issues per year
4. ISSN 0032-3640
5. Advertising, book reviews, charts, illustrations, trade literature, index
6. Cahners Publishing Company
   Division of Reed Elsevier, Inc.
   1350 E. Touhy Avenue
   Box 5080
   Des Plaines, IL 60017-5080

Preventive Medicine

1. Ernst L. Wynder
2. 1972
3. Bimonthly
4. ISSN 0091-7435
5. Index
6. Academic Press, Inc.
   Journal Division
   525 B Street, Suite 1900
   San Diego, CA 92101-4495

Professional Safety

1. Neal Lorenzi
2. 1956
3. Monthly
4. ISSN 0099-0027
5. Advertising, book reviews, charts, illustrations, trade literature, index
6. American Society of Safety Engineers

1800 E. Oakton Street
Des Plaines, IL 60018-2187

Public Health Reports

1. Dr. Anthony Robbins
2. 1878
3. Bimonthly
4. ISSN 0090-2918
5. Bibliographies, charts, illustrations, statistics, index
6. U.S. Public Health Service
   Department of Health and Human Services
   J.F.K. Federal Building, Room 1826
   Boston, MA 02203

Regulation

1. William A. Niskanen
2. 1977
3. 3 issues per year
4. ISSN 0147-0590
5. —
6. Cato Institute
   1000 Massachusetts Avenue NW
   Washington, DC 20001-5403

Science

1. Daniel Koshland
2. 1880
3. Weekly
4. ISSN 0036-8075
5. Book reviews, abstracts, bibliographies, illustrations, trade literature
6. American Association for the Advancement of Science
   1333 H Street NW
   Washington, DC 20005

Site Selection

1. Jack C. Lyne
2. 1956
3. 6 issues per year
4. ISSN 1041-3073

5. Advertising, cumulative index 1956-1994
6. Conway Data, Inc.
   40 Technology Park-Atlanta
   Norcross, GA 30092

Textile Chemist and Colorist

1. Susan H. Keesee
2. 1969
3. Monthly
4. ISSN 0040-490X
5. Advertising, book reviews, charts, illustrations, index, cumulative index
6. American Association of Textile Chemists and Colorists
   One Davis Drive
   Box 12215
   Research Triangle Park, NC 27709-2215

Textile Research Journal

1. Ludwig Rebenfeld
2. 1930
3. Monthly
4. ISSN 0040-5175
5. Book reviews, bibliographies, charts, illustrations, index
6. Textile Research Institute
   601 Prospect Avenue, Box 625
   Princeton, NJ 08542

Toxicology

1. Witschi, H. P. and K. J. Netter
2. 1973
3. 30 issues per year
4. ISSN 0300-483X
5. Advertising, charts, illustrations, index
6. Elsevier Science
   Regional Sales Office, Box 945
   New York, NY 10159-0945

Audiovisual Aids 5

The audiovisual aids listed in this chapter provide a wide range of information on indoor pollution. A graphic representation can often convey information more vividly than the written word. The films cover a wide variety of topics such as air pollution, physical pollutants, chemical toxicants, and noise pollution.

The following sources list audiovisual aid sources in English:

AFVA Evaluation, 1992. Fort Atkinson, WI: Highsmith Press, 1992.

Educational Film/Video Locator and University Media Centers and R. R. Bowker. 2 vols. 4th ed. New York: R. R. Bowker, 1990–1991.

Film and Video Finder. 3 vols. 4th ed. Medford, NJ: Plexus Publishing, 1994–1995.

Video Rating Guide for Libraries. Santa Barbara, CA: ABC/CLIO, various years.

The Video Source Book. 2 vols. and supp. 15th ed. Detroit: Gale Research Inc., 1994.

The following data are provided for each film:

Title of film
Distributor
Phone and fax numbers if available

Date of film

Description

Air Pollution

Air Pollution—Indoor
Films for the Humanities and Sciences
P.O. Box 2053
Princeton, NJ 08543-2053
Color, 26 minutes, sound, three-quarter-inch video, half-inch Beta/VHS, n.d.

Looks at the problem of indoor air pollution. Examines home improvements such as painting that cause pollutants and thermal insulation and waterproofing devices that trap pollutants.

Ozone

The Ozone Story
Educational Media Center
University of Utah
Salt Lake, UT 84112
Color, 15 minutes, sound, 1/2-inch VHS, n.d.

Explains that ozone is one of the most poisonous gases known to humans but necessary for their existence. Presents results of research at York University on the effect of fluorocarbon on ozone, the effect of ultraviolet light in producing sunburn and skin cancer, and ozone's effect on human beings.

Radon

Radon
Films for the Humanities, Inc.
P.O. Box 2053
Princeton, NJ 08540
Color, 26 minutes, sound, three-quarter-inch U-matic, half-inch VHS, 1988.

Considers the possible health implications of radon pollution. Methods that homeowners can use to detect radon are discussed; what can be done to minimize the radon hazard in homes.

Radon: A Homeowner's Guide
Acorns Media Publishing
7910 Woodmont Avenue
Suite 350
Bethesda, MD 20814
Color, 25 minutes, sound, half-inch VHS, 1987.

Tells everything you need to know about radon gas—what it is, its effect on your health, how to test for it, and how to interpret the test results. Also looks at remedies for radon problems.

Radon Free
Syber Vision
7133 Koll Center Parkway
Pleasanton, CA 94566
Color, 35 minutes, sound, half-inch VHS, 1990.

Shows how to detect radon poison in homes and how to remove it.

Tobacco Smoking

Smoking: A Research Update
University of Wyoming
Audio Visual Services
University Station, P.O. Box 3273
Laramie, WY 82071
Color, 27 minutes, sound, half-inch VHS, 1984.

Provides definitive answers to questions raised about smoking, including the effects of secondhand smoke.

Smoking against Your Will
Access Network-Alberta Educational Communications Corp.
3720-76 Avenue
Edmonton, Alberta
Canada T6B 2N9
Color, 30 minutes, sound, three-quarter-inch video, 1985.

Examines current medical research on the effect of the harmful ingredients absorbed in the bloodstream of those near smokers. Shows dangers of passive tobacco smoke.

Tobacco—Or Am I Just Blowing Smoke?
Media Guild
11722 Sorrento Valley Road
Suite E
San Diego, CA 92121
Color, 20 minutes, sound, half-inch VHS, three-quarter-inch U-matic, 1995.

Discusses smoking issues, especially health hazards of smoking to the smoker and those around the smoker; banning smoking in public places and businesses.

The Tobacco Problem: What Do You Think?
Encyclopedia Britannica Educational Corporation
310 S. Michigan Avenue
Chicago, IL 60604
Color, 17 minutes, sound, 16 mm, 1972.

Shows congressional hearings concerning the use of labels warning against smoking, rebuttal from consumers and tobacco companies, and debate by teenagers on the pros and cons of smoking. Asks the question: How would prohibition of smoking affect the national economy?

Lead Poisoning

Lead Paint Poisoning
U.S. Department of Commerce
National Bureau of Standards
Connecticut Avenue and Van Ness
Washington, DC 20234
Color, 5 minutes, sound, 16 mm film optical sound, 1972.

Looks at the lead paint problem and some of the steps taken by the federal government, especially the Department of Housing and Urban Development and the National Bureau of Standards, to overcome it.

Lead Poisoning
Films for the Humanities and Sciences
P.O. Box 2053
Princeton, NJ 08543-2053

Color, 26 minutes, sound, three-quarter-inch video, half-inch Beta/VHS, n.d.

Takes a look at the health hazards of lead. Examines what is being done to reduce and prevent lead contamination in air, water, and soil. Examines advances made in the treatment of people with lead poisoning.

Lead Poisoning
West Glen Communications
1430 Broadway
New York, NY 10018-3396
Color, 1 minute, sound, 16 mm film optical sound, 1973.

Warns parents with children who live in older houses and apartments of the dangers of lead poisoning. Shows children playing in areas where there is cracked and peeling paint and plaster.

Lead Poisoning Could Strike Your Child
Mar/Chuck Film Industries
P.O. Box 61
Mount Prospect, IL 60056
Color, 22 minutes, sound, 16 mm, 1980.

This film shows the most common sources of lead poisoning, the symptoms in children that reveal how children are affected by lead poisoning after they have eaten lead-contaminated paint, and the medical treatment of children who are affected by lead poisoning. The film concludes with the steps that can be taken, both temporarily and permanently, to remove lead contaminants from walls, ceilings, and woodwork.

Lead Poisoning: The Hidden Epidemic
Long Island Film Studios
P.O. Box 49403
Atlanta, GA 30359
Color, 9 minutes, sound, 16 mm, 1972.

Reveals the tragedy of lead poisoning in children. The film presents the sequence of events by first pointing out possible sources of lead in the home. This is followed by showing the symptoms of early and acute stages of lead poisoning, methods of detection, and means of treatment and prevention. Emphasis is placed on the need for frequent blood tests, prompt treatment, and elimination of lead sources in order to control a medical problem.

Lead Poisoning: What Everyone Needs to Know
Altschul Group Corporation
Health Division
1560 Sherman Avenue
Suite 100
Evanston, IL 60201
Color, 15 minutes, sound, half-inch VHS, 1993.

Discusses sources of lead, its health hazards, warning signs, and protection. Aimed primarily at parents.

Asbestos

Asbestos
Marshfield Video Network
Marshfield Clinic
Office of Medical Education
1000 North Oak Avenue
Marshfield, WI 54449-5777
Color, 41 minutes, sound, half-inch, VHS, 1979.

This tape offers detailed examples of asbestos forms and their effect on lungs.

Asbestos: A Lethal Legacy
Time-Life Film and Video
100 Eisenhower Drive
Paramus, NJ 07652
Color, 57 minutes, sound, 16 mm, three-quarter-inch U-matic, half-inch VHS, 1984.

Investigates the medical problems of asbestos contamination and examines the controversies regarding responsibility for health problems associated with asbestos. The question is raised as to how people can cope with the human risks of a useful but dangerous product. Part of the *Nova* public TV series.

Asbestos—A Matter of Time
U.S. Department of Interior Bureau of Mines
Graphic Services Section

4800 Forbes Avenue
Pittsburgh, PA 15213
Color, 20 minutes, sound, 16 mm film optical sound, 1959.

Covers the geological formation, production, and uses of this mineral fiber as well as dangers.

Asbestos Alert—Strategies for Safety and Health
AFL-CIO
Education Department
815 Sixteenth Street, NW
Washington, DC 20006
Color, 30 minutes, sound, half-inch VHS, 16 mm film optical sound, 1987.

Aims to educate workers on the hazards of asbestos and ways to prevent dangerous exposure to it. Aims to achieve safe working conditions with asbestos.

Asbestos-Hazard Awareness Training
Industrial Training Systems Corporation
9 East Stow Road
Marlton, NJ 08053-9990
Color, 48 minutes, sound, 1990.

Trains school custodians and maintenance employees who come in incidental or direct contact with asbestos or asbestos-containing materials about the hazards of asbestos. Discusses safe work habits as related to EPA's Asbestos Hazard Emergency Response Act.

Asbestos-Playing It Safe
U.S. National Audiovisual Center
8700 Edgeworth Drive
Capitol Heights, MD 20743-3701
Color, 37 minutes, sound, 16 mm film optical, 1985.

Looks at asbestos and the harm it can cause. Shows where asbestos can be found in a building and how it can be removed safely.

Asbestos Safety
International Film Bureau
332 Michigan Avenue

Chicago, IL 60604
Color, 25 minutes, sound, half-inch VHS, 1987.

The properties of asbestos and health hazards are discussed. States permissible exposure levels and indicates the warning signs revealing contamination; use of protective equipment. Consultants include a construction worker, an industrial hygienist, an asbestos abatement consultant, and an occupational health expert.

Asbestos Safety in the Schools
AFL-CIO
Education Department
815 Sixteenth Street NW
Washington, DC 20006
Color, 28 minutes, sound, three-quarter-inch video, 1984.

Describes efforts of a Philadelphia teacher to organize parents, teachers, and school employees to remove asbestos from the schools.

Asbestos: The Way to Dusty Death
Coronet/MTI Film and Video
108 Wilmot Road
Deerfield, IL 60015
Color, 51 minutes, sound, 16 mm, 1978.

Considers the several types of cancer that can develop after only a half-day of contact with asbestos. The story of lethal contact is revealed by a young mother who had embraced her father after his return from the mines and who is now dying from asbestos-caused disease. Reveals that the dangers from asbestos were known by individuals but ignored by people in responsible positions, a practice that continues.

Asbestos—Understanding the Hazards; Asbestos—Small Scale Short Duration Operations
Industrial Training Systems Corporation
9 East Stow Road
Marlton, NJ 08053-9990
Color, 51 minutes, sound, n.d.

Two videos on asbestos safety. Tells about health hazards involved in working with asbestos and necessary precautions in understanding the hazards. Shows how to remove pipe insulation, replace gaskets, and pack pumps when materials containing asbestos are involved.

Doin' It Right—Asbestos Abatement
National Audio-Visual Center
General Service Administration
Reference Section
Washington, DC 20409
Color, 22 minutes, sound, 16 mm, 1977.

Although the health problems of the use of asbestos were known since the 1940s, it was used extensively in fireproofing, insulation, and decoration of schools and buildings until 1978. Demonstrates for workers and school officials successful techniques for reducing asbestos exposure.

Mercury

Mercury Poisoning: A World Peril
Sterling Education Films, Inc.
241 East 34th Street
New York, NY 10016
Color, 26 minutes, sound, half-inch VHS, 1985.

Shows how a group of scientists are attempting to find a solution to the highly hazardous and often fatal problem of mercury contamination in the workplace.

Energy

Energy Efficient Housing: New Homes: The Airtight House, No. 5; Existing Homes: Plugging the Holes, No. 8
Access Network
Program Services
295 Midpark Way SE
Calgary, Alberta
Canada T2X 2A8
Color, 28 minutes, sound, half-inch VHS, n.d.

The series of 12 programs outlines the entire spectrum of building houses from the energy-efficient viewpoint. Numbers 5 and 8 discuss the advantages of energy efficiency and the problem of air contamination.

Chemicals

Chemical Safety for General-Service Workers
Cornell University
Audio Visual Research Center
8 Research Park
Ithaca, NY 14850
Color, 15 minutes, sound, three-quarter-inch U-matic, half-inch VHS, 1985.

A guide to safety for people, including mechanics, processing workers, painters, and food workers, who work with any brand of chemical. Describes how chemicals enter and affect the body, symptoms of chemical exposure and injury, and acute and chronic toxicity. How to obtain and use information on chemical toxicity.

Chemical Safety for Laboratory Workers
Cornell University
Audio Visual Research Center
8 Research Park
Ithaca, NY 14830
Color, 31 minutes, sound, three-quarter-inch U-matic, half-inch VHS, 1985.

Provides fundamental information on toxicology and safety procedures. Intended for use with research and laboratory workers who use a variety of chemicals. Designed to meet the need for information on the "right-to-know" laws. Includes symptoms of chemical exposure and injury, acute and chronic toxicity, and emergency procedures. How to receive toxicity information and make informed decisions.

Chemicals—A Fact of Life
Intercollegiate Video Clearing House
P.O. Drawer 33000r
Miami, FL 33133
Color, 27 minutes, sound, half-inch or three-quarter-inch video, n.d.

Describes natural chemicals in the environment, body cells, and man-made chemicals. Shows safety-testing measures. Explains the law and positive contributions of chemicals.

Chemicals and Common Sense
Intl. Film Bureau, Inc.
332 S. Michigan Avenue
Chicago, IL 60604-4382
Color, 12 to 18 minutes, sound, half-inch or three-quarter-inch video, 1986.

Discusses the hazard communication program. Tells where chemicals are kept and gives information about them.

Chemicals Under Control
Education Research Foundation
P.O. Box 928
Irmo, SC 29063
Color, 17 minutes, sound, half-inch or three-quarter-inch video, n.d.

Gives general information for employees who need to know about handling chemicals safely. Helps companies to meet requirements of OSHA chemical standards.

Chemicals Under Control—
Orientation to Basic Chemical Safety
Intl. Tele-Film
47 Densley Avenue
Toronto, Ontario
Canada M6M 5A8
Color, 17 minutes, sound, three-quarter-inch video, n.d.

Explains chemical safety for employees. Discusses toxicity, reactivity, and hidden hazards from changing conditions and combinations. Illustrates the safety factor of housekeeping, cleanup, detecting the presence of hazardous chemicals, and personal protection devices.

Toxins

Toxics in the Workplace
New York State Education Department
Media Distribution Network
Room C-7-CEC
Empire State Plaza
Albany, NY 12230

Color, 30 minutes, sound, half-inch VHS, Beta, three-quarter-inch U-matic, 1984.

An electrical worker fights the Massachusetts Legislature over "right-to-know" legislation that calls for labeling toxic chemicals in a workplace.

Toxics and Poisons
Aims Media, Inc.
9710 DeSoto Avenue
Chatsworth, CA 91311-4409
Color, 12 minutes, sound, half-inch VHS, 1991.

Explains the serious effects of toxic substances on the human body and shows safety precautions to be used when working with toxic and poisonous substances. Explains the "right-to-know" laws protecting employees who work with or around hazardous materials.

Toxins in Our Lives
Ames School Publishers
P.O. Box 543
Blacklick, OH 43004-0543
Color, 31 minutes, sound, half-inch VHS, 35 mm film strip, cassette, 1989.

Discusses the problem of toxic waste as a small-scale everyday home issue.

Noise

Noise
Intl. Film Bureau, Inc.
332 S. Michigan Avenue
Chicago, IL 60604-4382
Color, 20 minutes, sound, half-inch or three-quarter-inch video, n.d.

Discusses sound and recommends ways to reduce levels of noise. Illustrates how soundwaves are transmitted through the earth and measured.

Noise
Coronet/ MTI Film and Video
108 Wilmot Road

Deerfield, IL 60015
Color, 9 minutes, sound, half-inch or three-quarter-inch video, 1971.

Encourages employees to wear protection devices to prevent premature deafness.

Noise Abatement
Society of Manufacturing Engineers
One SME Drive
P.O. Box 930
Dearborn, MI 48121-0930
Color, 45 minutes, sound, half-inch: VHS, three-quarter-inch U-matic, n.d.

Presents a step-by-step noise-control program that includes planning, measurement, surveys, and implementations. Describes a General Motors program that was developed to control noise exposure to workers.

Noise and Its Effect on Health
Film Fair Communications
10621 Magnolia Road
North Hollywood, CA 91601
Color, 20 minutes, sound, half-inch VHS, three-quarter-inch U-matic, 1973.

Describes the adverse effect of noise on health including slow and permanent hearing loss.

Noise-Polluting the Environment
Omega Productions
3929 N. Humboldt Blvd.
Milwaukee, WI 53212
Color, 14 minutes, sound, half-inch or three-quarter-inch video, 1971.

Looks at ways to alleviate noise to prevent psychological effects.

Noise Pollution
Learning Corporation of America
Simon and Schuster Communications
108 Wilmot Road
Deerfield, IL 60015
Color, 18 minutes, sound, 16 mm, 1967.

Studies the effect of noise pollution that is harmful to the ear, psychological damage, and physical damage to structures. Combines physics, biology, and meteorology to aid in understanding problems and solutions.

Noise: The New Pollutant
Indiana University
Audio-Visual Center
Bloomington, IN 47405
Color, 30 minutes, sound, 16 mm, 1972.

Dr. Vern O. Knudsen, acoustical physicist at UCLA, demonstrates that sound is caused by differences in air pressure. The film reviews research into the harmful physiological and psychological effects of excessive noise on human beings.

Noise—The New Pollutant
Audio Visual Center
Indiana University
Bloomington, IN 47405-5901
Black and white, 30 minutes, sound, 16 mm film optical sound, 1967.

Presents research projects into the harmful effect of noise on human beings.

Noise—You're in Control
Agency for Instructional Technology
Box A
Bloomington, IN 47402-0120
Color, 14–15 minutes, sound, half-inch or three-quarter-inch video, 1986.

Demonstrates effective types of noise control and ear protection on the job.

Noise—You're in Control
BNA Communications, Inc.
9439 Key West Avenue
Rockville, MD 20850-3396
Color, 15 minutes, three-quarter-inch video, 1982.

Educates workers on effects of noise on hearing and the types of protection available and results if not worn.

Glossary

Indoor Air Pollution—General

abatement Actions to lessen or reduce the risk associated with contaminants in buildings.

absorption A process in which gases are transferred into a liquid or solid in which they dissolve.

adiabatic process A humidification process in which no heat is added to or taken away from the atmosphere.

adsorption A process by which negatively charged contaminants are attracted to a positively charged surface where they can be removed by scraping.

aerosol liquid Solid articles suspended in air that are sufficiently fine (0.01–100 mm) to remain suspended for an extended period of time.

air change rate Average number of times within one hour that indoor air is replaced by fresh outdoor air.

air cleaner A device that removes pollutants from indoor air; includes chemical filtration and electrostatic precipitation.

air-handling unit A part of a heating, ventilation, and air-conditioning system that moves and may also clean, heat, or cool air.

air quality standard A government-mandated regulation that specifies the maximum contaminant concentration beyond which health risks are unacceptable.

algae One of many plants of a subdivision of thallophyres found in both freshwater and salt water, including pond scums, kelps, and some seaweeds.

allergy Excess sensitivity to certain antigen substances such as food, drugs, pollen, or heat or cold, producing hay fever, asthma, or hives.

alpha energy Heat released when an alpha particle decays on a substance, such as a lung.

alpha particle Positively charged subatomic particle emitted from an atomic nucleus in the process of radioactive decay.

ambient air The total air mass; sometimes called outdoor air.

anemometer An instrument to measure the rate of air flow.

antigen A substance that produces an antibody when introduced into blood or tissue.

antimicrobial An agent that kills microbial growth.

aquifer A porous underground formation that contains groundwater.

arachnid Any member of a class of wingless arthropods with segmented bodies, including spiders, mites, and scorpions.

asbestos A naturally occurring group of fibrous minerals that is resistant to friction and corrosion, impervious to heat, and sound-absorbent. Chrysolite asbestos is used in 95 percent of the 3,000 commercial products that contain asbestos. Airborne asbestos fibers are highly toxic.

asbestos fibers Fibers with a length greater than 5 mm.

backdrafting Reverse airflow down flues, or chimneys that permit exhaust gases from combustion appliances to return indoors. Caused by negative air pressure. Generally associated with tightly sealed houses.

bacteria One-celled organisms that are members of the *Protista,* a biological classification.

bakeout A technique for reducing emissions from new buildings in which temperatures are raised (90° or higher) for several days to enhance emissions of volatile organic compounds from the new materials. Ventilation needs to be maintained at full capacity to exhaust the pollutants.

becquerel (Bq) The official international unit of radioactivity (3.70 x 10^{10} disintegration per second); has replaced use of the curie (Ci) in the international system.

binder A component of an adhesive composition that provides an adhesive force that holds two materials together.

bioaerosols Airborne microbial contaminants including viruses, bacteria, fungi, algae, and protozoa.

biocontaminants Contaminants that are either life-forms (e.g., molds of the genera *Aspergillus*) or are derived from living matter (e.g., rodent droppings).

bivalent Having a valence of two; combining with two atoms of a monovalent element or radical.

building envelope The outside of a building including walls, windows, doors, and roofs.

building related illness (BRI) A diagnosable illness with identifiable symptoms where cause can be directly traced to airborne contaminants within a building (e.g., Legionnaires' disease).

carcinogen Any substance capable of causing cancer.

ceiling plenum The space between the suspended and structural ceiling.

combustion by-products Any gas, including polynuclear aromatic hydrocarbons, carbon monoxide, and nitrogen dioxide, that is produced by the burning of wood or fossil fuels.

comfort zone The range in temperature and relative humidity in which 80 percent of the building occupants are comfortable.

concentration of pollution The amount of contaminants, particles, or gases in a given volume of ambient air.

condensation The precipitation of airborne water vapor.

condenser A heat-exchange instrument that removes vapor and heat from the atmosphere.

conditioned air Air that has been heated, cooled, humidified, or dehumidified in order to create a physical condition indoors.

contaminant An unwanted air constituent that may or may not be associated with or cause adverse health or discomfort.

contaminant exposure The contact of a pollutant with a susceptible part of the body.

convective movement An air movement resulting from differences in pressure due to temperature contrasts.

cooling capacity The maximum rate at which a cooling instrument removes heat from the atmosphere.

cooling tower A heat-transfer instrument that cools warm water using airflow. In most systems, water passes through tubing, thereby exposing the water to the cooler outside air.

dander Small scales from human or animal hair, skin, or feathers.

decay rate The rate at which the concentration of a compound decreases.

dehumidifier A device that removes water vapor from air.

dew point The temperature at which water vapor in the atmosphere condenses into a liquid.

diffuser A grill or inlet in a ventilating and air-conditioning system to distribute the flow of air in a room.

dilution The reduction of airborne concentrations of contaminants through an increase in outdoor air supply.

dose A given amount of a substance that reaches a site in the body where it causes an effect.

drain pan A container located beneath cooling coils in an air handling system to collect water condensate.

duct A tube or passage in an HVAC system for conveying air at low pressure.

dust Earth or other matter of fine, dry particles so small that it is easily raised and carried by the wind.

economizer In an atmospheric context, a central option in a ventilation system that makes optimum use of outside air supply to cool a building.

effluent Treated water that is returned to streams, lakes, or aquifers from municipal sewage plants, industrial operations, and household septic tanks.

electromagnetic field (EMF) The nonionizing electronic and magnetic radiation generated by electrical power lines and appliances.

electrostatic air cleaner A device that uses an electrical charge to remove particles moving into an airstream.

emission rate A measure of the quantity of pollutants released into the air in a given amount of time.

emission standard A guideline or government regulation that specifies the maximum rate at which a pollutant can be released from a given source.

emissions The release of airborne pollutants from a contaminated source.

encapsulant The covering of a pollutant to prevent the emission of pollutants.

environmental tobacco smoke Combustion emissions (consisting of over 3,800 identifiable contaminants, including 43 known or suspected carcinogens) released either by persons who are smoking tobacco or exhaling tobacco smoke.

ergonomics The applied science that investigates the impact of the physical environment on human stress, comfort, and health, considering such factors as climate control, household furnishings, noise, and overcrowding.

eutrophication A process through which a body of water becomes richer in nutrients and lower in dissolved oxygen content, reducing the waterway's ability to support fish. Many areas have banned phosphate detergents, the source of the excess nutrients.

exfiltration The passage of interior air to the outdoors through openings in a building.

exhaust air Air removal from all space and discharged into the atmosphere.

exhaust ventilation Localized ventilation by means of a device such as a fan to remove a contaminant from a specific source such as a kitchen or bathroom.

fan A tool with rotating blades that moves air. There are three types of fans: axial moves airflow parallel to the axis; centrifugal draws air in axially and propels it radially; and tubeaxial, an axial fan enclosed in a cylinder.

fiberglass A manufactured mineral fiber made from spun glass.

filter A device or medium used to remove gases and/or particles from an airstream.

flushout The operation of a ventilation system at its maximum level to remove airborne emissions from newly installed furnishings, carpeting, and other items.

fumes Smoke, vapor, or gas formed during the combustion process.

fungi Unicellular or multicellular organisms embracing a large group of microflora, including molds, mildews, yeasts, mushrooms, rusts, and smuts.

gamma radiation High-energy protons emitted by an atomic nucleus accompanied by radioactive decay.

gas Transparent molecules that expand to occupy a space or enclosure completely and uniformly. Gas molecules measure less than 0.0001 microns.

groundwater Rainwater that collects in an aquifer and supplies 50 percent of the drinking water in the United States.

half-life The time required for one-half of a given quantity of a radioactive substance to undergo decay.

health hazard A situation that has the capability of producing adverse health or safety conditions for humans.

heat The thermal form of energy.

high efficiency particulate air (HEPA) filter A type of high-efficiency filter that traps particulates.

house dust mite Common microscopic household arachnid that feeds on skin scales. Frequently found in furniture and mattresses. Mites require a relative humidity greater than 45 percent to survive.

humidifier A device that adds water to air.

humidifier fever A respiratory disease with influenza-like symptoms. Also called air conditioner or ventilation fever. It is caused by exposure to toxins emitted by microorganisms that become established in air conditioners or humidifiers.

hypersensitivity pneumonitis A group of respiratory diseases that involves inflammation of the lungs; caused by exposure to biological agents.

indoor air quality (IAQ) Characteristics of air inside a building.

infiltration The passage of outdoor air into a building through doors, windows, and cracks.

isothermal process A humidification process that adds moisture already converted to water vapor directly into the airstream.

isotope One of two or more types of atoms having the same atomic number (the number of protons) but a different mass number.

latent period The time elapsed between initial contact with a contaminant and its detection, often used in the context of fevers or cancer.

Legionnaires' disease A multisystem disease caused by *Legionella pneumophila* bacteria that affects not only the lungs but also the gastrointestinal tract, the kidneys, and the nervous system.

load A term referring to quantity of energy required per unit of time.

louver A covering for an opening that permits a flow of air while inhibiting the flow of water.

microbes Microorganisms known as bacteria, viruses, mold, fungi, spores, and pollen.

mildew A superficial covering of damp organic surfaces with fungi.

mist Suspended liquid droplets in the air, generated by condensation of moisture out of the atmosphere.

mitigation A procedure to eliminate an indoor air problem, either through ventilation control, exposure reduction, source control, or air-cleaning.

modifier A chemically inert ingredient added to an adhesive material that changes its properties but does not react chemically with the binder.

mol The molecular weight of a substance expressed in grams.

molar Characterizing a solution made up of one mol of solute per one liter of solution.

mold A fungal infestation that causes disintegration of a substance.

monitoring A process for assessing exposure to specific contaminants.

natural ventilation The movement of outdoor air into buildings due to pressure differences.

odor A chemical characteristic that is detected by the olfactory epithelium.

off-gassing The release of gases such as volatile organic compounds from materials after the manufacturing process is completed.

organism A living body ranging from a single-celled bacteria to humans and consisting of different parts, each specializing in a specific function.

outdoor air Ambient air drawn into a building that has not been previously circulated indoors.

out-gassing The emission of gases such as formaldehyde from building materials into the air due to chemical changes.

oxidation A reaction in which oxygen combines with another substance.

ozone (O_3) A highly reactive trivalent form of oxygen. Ozone exposure can create mucous membrane irritation and damage to the pulmonary system.

particulates Solid or liquid airborne matter that is larger than a single molecule but smaller than 0.5 mm.

permeability The capacity of a porous rock, solid, or sediment to transmit a liquid or gas without change through the medium.

permissible exposure limit (PEL) Air contaminant standard prepared by the Office of Safety and Health Agency (OSHA).

picocuries A trillionth of a curie; a curie is equal to 37 billion disintegrations of radioactivity per second.

plenum An enclosure of a heat, ventilation, and air-conditioning system that collects air at the beginning or ending of a duct system. Plenums are usually located above ceilings, below floors, or between walls.

pollutant A contaminant found in concentrations high enough to adversely affect human health or the environment. Often used synonymously with contaminant.

pollutant pathway A route of entry of an airborne contaminant from a source location into an area of human contact through architectural or mechanical openings in a building.

Pontiac fever A form of *Legionellosis* that is much milder than Legionnaires' disease. It has an incubation period of two to three days and attacks 90 percent of those infected. Pontiac fever is rarely reported, for it has symptoms similar to the flu.

protista All unicell animals and plants.

psychrometer An instrument to determine relative humidity by measuring both wet and dry bulb temperature.

radioactive decay The spontaneous transformation of a nuclide into one or more different nuclides accompanied by either the emission of energy and/or particles from the nucleus or the capture of an orbital electron by the nucleus.

radionuclide Any naturally occurring or artificially produced chemical element or isotope that undergoes spontaneous disintegration.

radon daughters The intermediate members in the decay series of radon. The term applies to the short-lived decay products.

recirculated air Indoor air that passes through the HVAC system to be heated or cooled and is then recirculated through the building.

relative humidity The percentage of moisture in the air relative to the amount it could hold if saturated at the same temperature.

respirable suspended particles (RSP) Inhalable particulate matter of less than 10 mm in diameter.

return air The air returned to occupied space after being altered, such as by raising or lowering temperature.

sealant Any material used to prevent the passage of liquid or gas through any opening.

sealer Any material such as plastic that prevents the passage of a gas or liquid through it.

sick building syndrome A set of symptoms shared by many building occupants that are not associated with a specific disease or caused by a known pollutant but that diminish or disappear entirely when occupants leave the building.

smoke Airborne particles (either solid or liquid) that are produced by incomplete combustion of carbon-containing materials.

soil gases Gases that enter a building from the ground (e.g., radon).

solute A dissolved substance.

solvent A substance that can dissolve another substance. Commonly used chemical solvents are trichloro-ethylene, xylene, trichloroethane, toluene, and benzene.

source control A preventive strategy for reducing airborne pollutants through the removal of the material or activity producing the contaminant.

stack effect Because heat rises in a building, the air pressure in the upper floors is lower than in the ground floor. This creates an air movement from the lower floor to the higher floors. As a result, normal ventilation is disrupted.

sublimation To convert a solid substance, by heat, into a vapor, which upon cooling condenses again to solid form without apparent liquefaction.

surface water Water in rivers, lakes, ponds, and reservoirs.

temperature The thermal state of a substance.

threshold The contaminant dose or exposure level that has an adverse effect on health.

threshold limit value standards The American Council of Governmental Industrial Hygienists-recommended guidelines for a contaminant exposure limit represented in terms of exposure over a workday (eight hours) or workweek (40 hours).

toxicity The nature and degree of a contaminant's adverse effects on living organisms.

tracer gas An inert compound that is a rare constituent of indoor air, such as sulfur hexafluoride, which is released into buildings' air and monitored qualitatively and/or quantitatively to determine airflow characteristics. Applications include containment pathway detection, infiltration, and ventilation efficiency.

vapor The gaseous state of a substance that originates from a solid or liquid.

ventilation rate The rate at which fresh outdoor air is introduced to replace air in a building through natural or forced outdoor air exchange. Measured as air changes per hour (ACH) or cubic feet per minute (cf/m).

ventilation standard The specification for the minimum rate of input of outdoor air into indoors to maintain quality air conditions.

Chemical Compounds

acetaldehyde, CH_3CHO A water-soluble and colorless liquid that evaporates rapidly and has a pungent odor, used mainly in organic synthesis and in silvering mirrors.

acetate A salt formed by the union of acetic acid, CH_3COOH with a base. The common forms are ethyl and vinyl acetates.

acetone, $(CH_3)_2CO$ A volatile chemical compound used as a solvent for paints and varnishes, rubbers, and plastics and in organic synthesis.

acrolein, CH_2CHCHO The volatile liquid aldehyde having an acrid odor, obtained by the oxidation of allyl alcohol, used in organic synthesis.

alcohols Any of a class of chemical compounds derived from hydrocarbons by replacing one or more of the hydrocarbon atoms with an equal number of hydroxyl radicals, ethyl alcohol C_2H_6OH, the alcohol of commerce and medicine. Other alcohols are benzyl, ethyl, and isopropyl alcohol.

aldehyde One of a group of highly reactive organic compounds that contain the common group CHO, such as formaldehyde (HCHO) and acetaldehyde.

aliphatic Organic compounds that have only an open chain formed by carbon atoms.

aliphatic hydrocarbons A class of organic chemicals in which the carbon atoms are attached to each other in a straight chain or branch chain. Some aliphatic compounds have been identified as carcinogens, including propane, butane, isobutane, n-decane, and alkanes.

alkane One of a group of saturated aliphatic hydrocarbons present in many waxes, polishes, and lubricants.

amines Any of the derivative compounds of ammonia in which the hydrogen atoms are replaced by one or more organic hydrocarbon radicals. Chemical compounds added to steam boiler water to inhibit corrosion. Humidification systems using boiler steam may transfer these potentially dangerous chemicals to the airstream. Amines include ethylene diamine, triethanolamine, and isopropanolamine.

amnconia, NH_3 A colorless, pungent, suffocating alkali gas. Very soluble in water, forming ammonia water or aqueous ammonia.

a-pinene, $C_{10}H_{16}$ A liquid isomeric terpene forming the principal constituent of oil of turpentine; also occurs in other essential oils.

aromatic hydrocarbons A large class of highly volatile hydrocarbons containing a benzene ring, including such organic chemicals as toluene, tylene, ethylbenzene, and styrene. Examples of aromatic compounds include paints, adhesives, and solvents.

benzene, C_6H_6 A clear, colorless, aromatic liquid; the simplest aromatic hydrocarbon enacted from coal tar. Used as a solvent and intermediate in manufacturing organic chemicals. Determined by occupational studies to be a confirmed human (Class A) carcinogen. The principal indoor source of benzene is environmental tobacco smoke. Other sources include petroleum products such as gasoline, oil, and certain petroleum-based solvents.

butane, C_4H_{10} A colorless flammable aliphatic hydrocarbon gas produced from petroleum and natural gas, used as a refrigerant or solvent. Also isobutane.

chlordane, $C_{10}H_6Cl_6$ A colorless, odorless, viscous liquid used as an insecticide.

chloride A green-yellow gas highly irritating to human lungs; used to manufacture solvents, pesticides, plastics, bleaches, and many other products.

chloroform A volatile colorless liquid with a fragrant taste and smell, prepared by distilling a mixture of alcohol, water, and chloride of lime; or a by-product of the chlorination of methane used to manufacture such products as propellants, plastics, and dyes.

esters One of a class of organic compounds corresponding to inorganic salts; formed by the reaction of an alcohol with an acid and the elimination of water; includes ethyl acetate and alkyl ethoxylate.

ethers, $(C_2H_5)_2O$ A highly volatile and inflammable colorless liquid, used as a solvent. Other common forms are ethyl methyl and butyl ether.

ethyl A univalent hydrocarbon radical used to produce ethyl acetate and ethyl alcohol.

ethyl acetate A colorless, flammable, fragrant liquid used as a paint and lacquer solvent and in perfume and organic synthesis.

ethyl alcohol A colorless, limpid, volatile liquid used as a solvent; an intermediate in organic synthesis commonly known as grain alcohol.

ethylene glycol (ethylene alcohol), CH_2OHCH_2OH A clear, colorless, syrupy liquid with a sweet taste, used in organic synthesis and manufacturing of plastics, brake fluid, and radiator antifreeze.

formaldehyde (HCHO) An odorous, poisonous, volatile organic compound that contains oxygen, carbon, and hydrogen. Used as a solution in the manufacturing of synthetic resins and other organic compounds.

glycol Any of a group of alcohols of a similar type; ethylene glycol.

glycol acid A crystalline, colorless compound, $CH_2OHCOOH$, obtained by oxidation of glycol; used in leather and textile dyeing.

halogenated hydrocarbons An organic compound that has had one or more hydrogen atoms replaced by fluorine, chlorine, bromine, or iodine;

includes tetrachlorethane, methylene chloride, 1,1,1-trichloroethane, and 1.4 dichlorobenzene.

hydrocarbon Any one of the compounds consisting of hydrogen and carbon atoms only, such as benzene or methane.

kerosene A liquid hydrocarbon distilled from coal, bitumen, or petroleum; used as a fuel for stoves and as a cleaning agent or solvent.

ketones A class of organic compounds, such as acetone, each consisting of a carbonyl group united to one bivalent or two monovalent hydrocarbon radicals. Used as a solvent in manufacturing. Other forms of ketones are methyl isobutyl, methyl ethyl ketone, and acetone.

limonene A liquid terpene.

methanol, CH_3OH A flammable, water-soluble, volatile, highly toxic liquid alcohol used in chemical synthesis; used as a fuel solvent and denaturant for ethyl alcohol.

methyl, CH_3 A univalent hydrocarbon radical.

monovalent An elementary substance, one atom of which enters into combination with a single atom of another elementary substance.

n-decane, $C_{10}H_{22}$ Any one of several liquid hydrocarbons of the methane series, which occur in isometric forms.

n-hexane One of the isomeric hydrocarbons, volatile and liquid in nature, produced from petroleum.

pesticides A chemical used to kill or control living organisms. Pesticides include insecticides, herbicides, fungicides, rodenticides, antimicrobial agents, and plant growth regulators.

4-Phenylcyclohexane (4-PC) An odiferous compound that is a by-product of the manufacture of styrenebutadiene latex backing of some carpets.

polycyclic aromatic hydrocarbons (PAHs) A group of complex organic substances with four or more ring structures. Generally associated with certain combustion processes such as wood burning, tobacco smoking, and cooking; also known as polynuclear aromatic hydrocarbons or polynuclear hydrocarbons.

polymer The union of two or more molecules of a compound to form a more complex compound with a higher molecular weight; the conversion of one compound into another by such a process.

polyurethane A polymer with a light, foamy texture resulting from the entrapment of carbon dioxide in pores during production; used in the manufacture of padding and various resins.

radical An atom or group of atoms regarded as an important constituent of the molecule of a given compound and which remains unchanged during certain reactions.

resin A nonvolatile solid or semisolid organic material, usually of high molecular weight. Soluble in most organic solvents but not in water. The film-forming component of a paint or varnish; used in making plastics or adhesives.

styrene, $C_6H_5CH = CH_2$ A colorless liquid hydrocarbon with an aromatic odor; used in making synthetic rubber, plastics, and resins.

terpene, $C_{10}H_{16}$ Any of certain unsaturated hydrocarbons occurring as essential oils in many plants.

tetrachloride Any compound having four atoms of chlorine.

toluene, $C_6H_5CH_3$ A colorless, flammable, mobile liquid hydrocarbon obtained from coal tar and petroleum and used in making explosives and dyes and as a solvent.

urethane, $NH_2CO_2C_2H_5$ White crystalline substance found in medicines and fungicides.

vinyl, CH_2CH The univalent radical in many compounds to produce vinyl acetate; a type of plastic made from these resins.

volatile organic compounds (VOCs) A class of chemical components that contain one or more carbon atoms and are volatile at room temperature and normal atmospheric pressure. Indoor VOCs are generated by such sources as tobacco smoke, building materials, furnishings, cleaning products, solvents, polishes, cosmetics, deodorizers, and office supplies.

xylenes, $C_6H_4(CH_2)_2$ Any of the three isomeric hydrocarbons of the benzene series, occurring as oily, colorless, liquids obtained chiefly from coal tars and used mainly in the manufacturing of dyes.

Acronyms

ASC	Adhesive and Sealant Council
AWMA	Air and Waste Management Association
APCA	Air Pollution Control Association
ABIH	American Board of Industrial Hygiene
ACGIH	American Conference of Governmental Industrial Hygienists
ACEC	American Consulting Engineers Council
AFA	American Floorcovering Association
AIHA	American Industrial Hygiene Association
AIA	American Institute of Architects
ANSI	American National Standards Institute
APHA	American Public Health Association
ASHRAE	American Society of Heating, Refrigerating, and Air-Conditioning Engineers

ASTM	American Society for Testing and Materials
AWC	American Wood Council
AIA	Asbestos Information Association
AEE	Association of Energy Engineers
AWCII	Association of the Wall and Ceiling Industries International
BOCA	Building Officials and Code Administration
BOMA	Building Owners and Managers Association International
CRI	Carpet and Rug Institute
CSI	Construction Specification Institute
CPSC	Consumer Product Safety Commission
DHHS	Department of Health and Human Services
EPA	Environmental Protection Agency
FHA	Federal Housing Administration
GSA	General Services Administration
HUD	(Department of) Housing and Urban Development
IESNA	Illuminating Engineering Society of North America
NASA	National Aeronautics and Space Administration
NBS	National Bureau of Standards
NIOSH	National Institute of Occupational Safety and Health
NPCA	National Paint and Coatings Association
NPA	National Particleboard Association
NPCA	National Pest Control Association
OSHA	Occupational Safety and Health Administration
TIMATS	Thermal Insulation Manufacturers Association Technical Services
WHO	World Health Organization

Measurements

bq	becquerel
bq/m^3	becquerels per cubic meter
cfm	cubic feet per minute
dB	decibel
dBA	A weighted decibel
f/cm^3	fibers per cubic centimeter
F/m^3	fibers per cubic meter
fpm	feet per minute
g	gram
Hz	hertz—one complete cycle in a sound wave which is propagated through air
μg/m^3	micrograms per cubic meter
μg/cm^2	micrograms per square centimeter
μg/dl	micrograms per deciliter
μg/ft^2	micrograms per square foot
μ	micron; a unit of linear measure equal to one-millionth of a meter or one-thousandth of a millimeter

mps	meters per second
m/s	meters per second
mg/h	milligrams per hour
mg/m^2	milligrams per square meter
mg/m^3	milligrams per cubic meter
ml/mn	milliliters per minute
ppb	parts per billion
ppm	parts per million
pci/l	picocuries per liter
WLs	working levels

Testing Information

Consumer Product Safety Commission
Office of the Secretary
5401 Westbard Avenue
Bethesda, MD 20207
(800) 638-2772

Consumer Product Safety Commission provides lists of local laboratories that will test for a range of contaminants.

Environmental Protection Agency
Toxic Substances Control Assistance Office
401 M Street SW
Washington, DC 20460
(202) 544-1404 or (800) 835-6700

Provides a list of noncommercial laboratories that have been accredited to test for toxic substances.

Atmosphere

Ventilation Systems

American ALDES
Ventilation Corp.
4539 Northgate Court
Sarasota, FL 34234

DEC International
1919 South Stoughton Road
Madison, WI 53716
(800) 533-7533 or (608) 222-3484

Riehs and Riehs
501 George Street

New Bern, NC 28560
(919) 636-1615
Therma-Stor Products Group
Box 8050
Madison, WI 53708
(608) 222-3484 or (800) 533-7533
FAX: (608) 222-9314

Air Testing Services and Equipment

BCM Engineers, Inc.
One Plymouth Meeting
Suite 506
Plymouth Meeting, PA 19462
(215) 825-3800
FAX: (215) 834-8236

Chem-Safe Laboratories
Box 546
Pullman, WA 99163
(509) 334-0922

Electro-Analytical Laboratories
7118 Industrial Park Boulevard
Mentor, OH 44060
(216) 951-3514

Environmental Testing and Technology
Box 369
Encinatas, CA 92024
(619) 436-5990

Lancaster Laboratories
2425 New Holland Pike
Lancaster, PA 17601
(717) 656-2301

RK Occupational and Environmental Analysis
616 Warren Street
Alpha, NJ 08865
(201) 454-6316

Safe Environments Home and Office Testing Laboratories
Box 489
San Leandro, CA 94577
(415) 843-6042
FAX: (415) 635-6730

Carbon Monoxide Detectors

Assay Technology
1070 East Meadow Circle
Palo Alto, CA 94330
(800) 833-1258 or (415) 424-9947

Blue Sky Testing Labs
8655 39th Avenue, South
Seattle, WA 98118
(206) 721-2583

Lab Safety Supply
Box 1368
Janesville, WI 53547
(800) 356-0783

Neotronics N.A.
2144 Hilton Drive, NW
Gainsville, GA 30503
(800) 535-0606 or (404) 535-0600 in Georgia
FAX: (404) 532-9282

Quantum Group, Inc.
Box 210347
San Diego, CA 92121
(800) 432-5599 or (619) 457-3048 in California

Sensidyne
12345 Starkey Road
Largo, FL 34643
(800) 451-9444 or (813) 530-3602 in Florida
FAX: (813) 539-0050

Nitrogen Dioxide Test Kits

MDA Scientific
405 Barclay Road
Lincolnshire, IL 60069
(800) 323-2000

Air Purifiers

Electrostatic Air Cleaners

Air Control Industries
213 McLemore Street
Nashville, TN 37203
(615) 242-3448

Air Quality Engineering
3340 Winpark Drive
Minneapolis, MN 55427
(800) 328-0787 or (612) 544-4426 in Minnesota
FAX: (612) 544-4013

The Allergy Store
Box 2555
Sebastopol, CA 95473
(800) 824-7163 or (800) 950-6202 in California

The Environmental Network
6015 East Commerce, Suite 430
Irving, TX 75063
(214) 550-0808

Honeywell
1985 Douglas Drive, North
Golden Valley, MN 55422
(612) 542-6723

Newtron Products
3874 Virginia Avenue
Cincinnati, OH 45227
(800) 543-9149 or (513) 561-7373 in Ohio

Permatron Corporation
10134 Pacific Avenue
Franklin Park, IL 60131
(800) 882-8012
(312) 678-0314 in Illinois

Summit Hill Laboratories
Navesink, NJ 07752
(201) 291-3600

Tectronic Products
6500 Badgley Road
East Syracuse, NY 13057
(315) 463-0240
FAX: (315) 437-7290

Universal Air Precipitator Corporation
150 McCully Road
Monroeville, PA 15146
(412) 372-0706

HEPA Air Cleaners and Activated Carbon Filters

Airguard Industries
Box 32578
Louisville, KY 40232
(502) 969-2304
FAX: (502) 969-2579

Allergen Air Filter Corporation
5205 Ashbrook
Houston, TX 77081
(800) 333-8880

Brian Pure Aire
9307 State Route 43
Streetboro, OH 44241
(800) 458-5200 or (216) 626-5400 in Ohio

Cameron-Yakima, Inc.
Box 1554
Yakima, WA 98907
(509) 452-6605

Dust Free
Box 454
Royse City, TX 75089
(800) 441-1107 or (214) 635-9565 in Texas

Envirocaire
747 Bowman Avenue
Hagerstown, MD 21740
(800) 332-1110 or (301) 797-9700 in Maryland

E. L. Foust
Box 105
Elmhurst, IL 60125
(800) 225-9549

Mason Engineering and Designing
242 West Devon Avenue
Bensenville, IL 60106
(312) 595-5000

Research Products
1015 East Washington Avenue
Madison, WI 53703
(800) 356-9652 or (608) 257-8801 in Wisconsin

3M Filtration Products
76-1W, 3M Center
St. Paul, MN 55144
(612) 733-1110

Vitaire
PO Box 88
Elmhurst Annex, NY 11380
(800) 447-4344 or (201) 473-2244 in New York

HEPA Vacuum Cleaners

Critical Vacuum Systems
6862 Flying Cloud Drive
Eden Prairie, MN 55344
(800) 328-8322 ext. 582 or (612) 829-0836 in Minnesota

Euroclean
907 West Irving Park Road
Itasca, IL 60143
(800) 323-3553 or (312) 773-2111 in Illinois

Healthmore, Inc.
3500 Payne
Cleveland, OH 44114
(216) 432-1990 or (800) 344-1840

NFE International Limited
302 Beeline Drive
Bensenville, IL 60106
(800) 752-2400 or (312) 350-1110 in Illinois
FAX: (312) 350-1033

Nilfisk of America
300 Technology Drive
Malvern, PA 19355
(215) 647-6420

Rexair, Inc.
3221 West Big Beaver
Suite 200
Troy, MI 48084
(313) 643-7222

Vactagon Pneumatic Systems
25 Power Avenue
Wayne, NJ 07470
(201) 942-2500

Ion-Exchange Air Cleaners

Air Ion Devices
P.O. Box 5009
Novato, CA 94948-5009
(800) 388-4667

Air-to-Air Heat Exchangers

Airxchange
401 V.F.W. Drive
Rockland, MA 02370
(617) 871-4816
FAX: (617) 871-3029

Amerix
15 South 15th Street
Fargo, ND 58103
(701) 232-4116

Berner Air Products
P.O. Box 5410
New Castle, PA 16105
(800) 852-5015 or (412) 658-3551 in Pennsylvania

Des Champs Laboratories
Box 440
East Hanover, NJ 07936
(201) 884-1460
FAX: (201) 994-4660

EER Products
3536 East 28th Street
Minneapolis, MN 55406
(612) 721-4231
FAX: (612) 721-6303

Engineering Development
4850 Northpark Drive
Colorado Springs, CO 80918
(719) 599-9080

Nutech Energy Systems
124 Newbold Court
London, Ontario N6E 1Z7
Canada
(519) 686-0797

Q-Dot Corporation
701 North First Street
Garland, TX 75040

Snappy A-D-P
PO Box 1168
Detroit Lakes, MI 56501
(218) 847-9258

Water

Water Hotline

Safe drinking water hotline:
(800) 426-4791 or (202) 382-5533 in Washington

The hotline is located in the Environmental Protection Agency and was established to explain the Safe Drinking Water Act and its enforcement to the public.

Water Testing Laboratories

The following private laboratories can evaluate tap water in private residences:

Brown and Caldwell Laboratories, (415) 428-2300 or (818) 795-7553
Montgomery Laboratories, (818) 796-9141
Wilson Laboratories, (800) 255-7912; Kansas residents call (800) 432-7921
Suburban Water Testing Laboratories, (800) 433-6595; Pennsylvania residents call (800) 525-6464
Water Test Corporation, (800) 426-8378
National Testing Laboratories, (800) 458-3330

Testing for Chlorine

The following suppliers have testing equipment that allows you to find out whether chlorine is getting through your filter into finished water:

Chemetrics
Route 28
Calverton, VA 22016
(800) 356-3072 or (703) 788-9026 in Virginia

Hach Chemical Company
Box 389
Loveland, CO 80539
(800) 227-4224 or (303) 669-3050 in Colorado

LaMotte Chemical Products
Box 329

Chestertown, MD 21620
(800) 344-3100 or (301) 778-3100 in Maryland
FAX: (301) 778-6394

Taylor Chemicals
31 Loveton Circle
Sparks, MD 21152
(800) 638-4776
FAX: (301) 771-4291

Water Purification Devices

Activated Carbon Filters

Cuno
400 Research Parkway
Meriden, CN 06950

The Environmental Network
6015 East Commerce, Suite 430
Irving, TX 75063
(214) 550-0808

Multi-Pure Drinking Water Systems
9200 Deering Avenue
Chatsworth, CA 91311
(818) 341-7577

Reverse-Osmosis Filters

Aquathin Corporation
2800 West Cypress
Creek Boulevard
Fort Lauderdale, FL 33309
(305) 977-7997
FAX: (305) 978-6812

General Ecology
151 Sheree Boulevard
Lionville, PA 19353
(215) 363-7900
FAX: (215) 363-0412

Distillers

Dupont
Applied Technology Division
P.O. Box 110
Kennett Square, PA 19348

Durastill
875 Brookfield Parkway
Rosewell, GA 30075
(404) 993-7575

Jordan Chemical Company
PO Box 3164
Pikeville, KY 41501

Polar Bear Water Distillers
829 Lynnhaven Parkway, Suite 119
Virginia Beach, VA 23452
(800) 222-7188 or (800) 523-6388 in Virginia

Superstill Technology
888 Second Avenue
Redwood City, CA 94063
(415) 366-1133

Technical Services Department
3M Corporation
3M Center
St. Paul, MN 55144
(612) 773-1110

Chemicals

Formaldehyde Testing Laboratories

Air Quality Research
901 Grayson Street
Berkeley, CA 94710
(415) 644-2097

Assay Technology
1070 East Meadow Circle
Palo Alto, CA 94330
(800) 833-1258 or (415) 424-9947

Crystal Diagnostics
30 Commerce Way
Woburn, MA 01801
(617) 933-4114

3M
Occupational Health and Environmental Safety Division
3M Center, Building 220-3E-04

St. Paul, MN 55144
(612) 733-6486

Nontoxic Household Products and Building Materials

AFM Enterprises
1140 Stacy Court
Riverside, CA 92507
(714) 781-6860
FAX: (714) 781-6892

Cleansers, water-based paints, paint strippers, sealers to stop formaldehyde emissions, waxes, polishes, glues, shampoos

The Allergy Store
Box 2555
Sebastopol, CA 95473
(800) 824-7163 or (800) 950-6202 in California

Cleansers, skin-care products, shampoos

Baubiologie Hardware
207B 16th Street
Pacific Grove, CA 93950
(408) 372-8626

Soaps, cleansers, waxes, paints, glues

Beneficial Insectary
14751 Oak Run Road
Oak Run, CA 96069
(916) 472-3715

Parasites that kill houseflies before they hatch, reducing the need for home insecticides

Bon Ami
Faultless Starch Company
1025 West 8th Street
Kansas City, MO 64101
(816) 842-1230

Chlorine-free household cleanser; available in most stores

Dumond Chemicals
1501 Broadway
New York, NY 10036
(212) 869-6350

Lead-based paint removal system

Ecco Bella
6 Provost Square, Suite 602
Caldwell, NJ 07006
(201) 226-5799
FAX: (201) 226-0991

Cosmetics, household cleansers, safe pesticides

Livos Plant Chemistry
2641 Cerrillos Road
Santa Fe, NM 87501
(800) 621-2591 or (505) 988-9111

Paints, adhesives, shellacs, stains, waxes, art materials, household cleansers

Oakmont Industries
44 Oak Street
Newton Upper Falls, MA 02164
(800) 447-2229
FAX: (617) 899-8726

Nontoxic pesticides

Pace Chem Industries, Inc.
779 LaGrange Avenue
Newbury Park, CA 91320
(805) 496-6224

Paints, sealants

Shaklee Corporation
444 Market Street
San Francisco, CA 94111
(800) 544-8860
Sinan Company
Box 181
Suisun City, CA 94585
(707) 427-2325

Paints, waxes, soaps

The Sprout House
40 Railroad Street
Great Barrington, MA 01230
(413) 528-5200

Household cleansers

Sunrise Lane
780 Greenwich Street
New York, NY 10014
(212) 242-7014

Cosmetics, cleansers, soaps

Organic Chemicals Test Kits

Industrial Scientific Corporation
355 Steubenville Pike
Oakdale, PA 15071
(412) 758-4353

National Draeger, Inc.
Box 120
Pittsburgh, PA 15230
(412) 787-8383

Technical Services Department
3M Corporation
3M Center
St. Paul, MN 55144
(612) 773-1110

Radon Test Kits

Air Chek, Inc.
180 Glenn Bridge Road
Box 2000
Arden, NC 28704
(800) 257-2366 or (704) 684-0893 in North Carolina

American Radon Consultants
3282 Chalfont Road
Shaker Heights, OH 44120
(216) 921-8398

American Radon Corporation
8300 North Hayden Road, Suite 100
Scottsdale, AZ 85258
(602) 967-8029

Applied Technical Services
1190 Atlanta Industrial Drive
Marietta, GA 30066
(800) 451-3405 or (404) 432-1400 in Georgia

Nuclear Associates
100 Voice Road
Carle Place, NY 11514
(516) 741-6360

Radon Analytical Laboratories
8935 North Meridian Street, Suite 110
Indianapolis, IN 46260
(317) 843-0788

Radon Detection Services
Old York Road
Ringoes, NJ 08551
(201) 88-3080

Radon Testing Corporation of America
50 South Buchout
Irvington, NY 10533
(800) 457-2366 or (800) 537-7822 in New York

Asbestos

Asbestos Testing Laboratories

Alternative Ways
100 Essex Avenue
Bellmawr, NJ 08031
(800) 547-0101

Building Environmental Systems
3501 North MacArthur, Suite 400B
Irving, TX 75062
(800) 982-0030 or (214) 257-0787 in Texas

Laboratory Testing Services
75 Urban Avenue
Westbury, NY 11590
(800) 433-0008

Source Finders Information
P.O. Box 758
Mt. Laurel, NJ 08054
(609) 482-1151

Sources of Volatile Organic Compounds (VOCs) in Indoor Air

Adhesives

alcohols
aliphatic hydrocarbons
esters
ethers
halogenated hydrocarbons
ketones
organic nitrogen compounds

Building Materials

aldehydes
aliphatic hydrocarbons
aromatic hydrocarbons
esters
ethers
halogenated hydrocarbons
ketones

Cleaners and Waxes

alcohols
aldehydes
aliphatic hydrocarbons
aromatic hydrocarbons
esters
ethers
terpenes

Combustion Appliances

aldehydes
aliphatic hydrocarbons

Cosmetics/Personal Care Products

alcohols
aldehydes
esters
ethers
ketones

Furniture and Household Furnishings

aldehydes
aromatic hydrocarbons
esters
ethers
halogenated hydrocarbons

Heating, Ventilating, and Air-Conditioning Systems

aliphatic hydrocarbons

Hobby Supplies

alcohols
aldehydes
aliphatic hydrocarbons
amines
esters
ethers
halogenated hydrocarbons
ketones

Human and Biological Origins

alcohols
aldehydes
aliphatic hydrocarbons
aromatic hydrocarbons
ketones

Paints and Related Products

aliphatic hydrocarbons
aromatic hydrocarbons
alcohols
esters
ethers
halogenated hydrocarbons
ketones

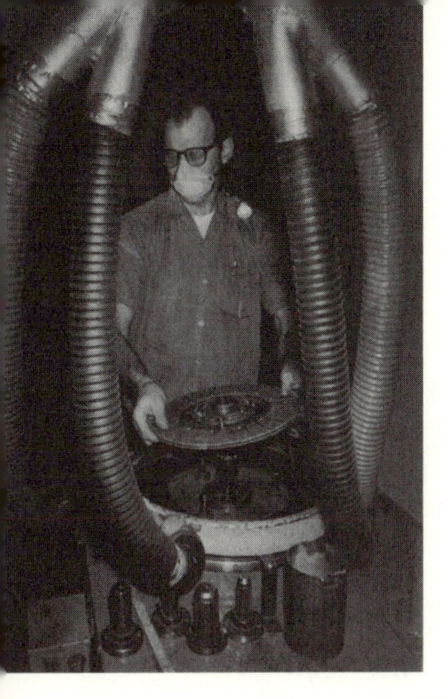

Index

E. Willard Miller is a professor of geography and associate dean of resident instruction (emeritus) at the Pennsylvania State University. He received his A.M. degree at the University of Nebraska and his Ph.D. from the Ohio State University. He is a fellow of the American Association for the Advancement of Science, the American Geographical Society, the National Council for Geographic Education, and the Explorers Club. In 1990, Dr. Miller received the Honors Award from the Association of American Geographers. He has published more than 100 professional journal articles and 30 books. He is listed in Who's Who in America and Who's Who in the World.

Ruby M. Miller is a map librarian (retired) at the Pattee Library at the Pennsylvania State University. She received her B.A. degree at Chatham College and did graduate work at the University of Pittsburgh. She established and developed the map collection at Penn State, and at her retirement the collection had over 250,000 maps and over 3,000 atlases. She is the co-author of 6 books and over 100 bibliographies. The Pennsylvania Geographical Society has honored her with its Distinguished Scholar Award. She is a member of the Association of American Geographers and the American Association of University Women.